Phenolic Compounds in Fruit Beverages

Special Issue Editor

António Manuel Jordão

MDPI • Basel • Beijing • Wuhan • Barcelona • Belgrade

MDPI

Special Issue Editor
António Manuel Jordão
Polytechnic Institute of Viseu (CI&DETS)
Chemistry Research Centre (CQ-VR)
Portugal

Editorial Office
MDPI
St. Alban-Anlage 66
Basel, Switzerland

This edition is a reprint of the Special Issue published online in the open access journal *Beverages* (ISSN 2306-5710) from 2017–2018 (available at: http://www.mdpi.com/journal/beverages/special_issues/phenolic-bev).

For citation purposes, cite each article independently as indicated on the article page online and as indicated below:

Lastname, F.M.; Lastname, F.M. Article title. *Journal Name* **Year**, *Article number*, page range.

First Edition 2018

ISBN 978-3-03842-985-2 (Pbk)
ISBN 978-3-03842-986-9 (PDF)

Table of Contents

About the Special Issue Editor

António Manuel Jordão is currently Assistant Professor of Oenology at Polytechnic Institute of Viseu (CI&DETS), Agrarian Higher School and member of the Chemistry Research Centre—Food and Wine Lab, Vila Real. He is graduated in Agro-Industrial Engineering and has a Master in Food Science and Technology from Technical University of Lisbon. In 2006, obtained a PhD in Agro-Industrial Engineering at Agronomy Higher Institute of Technical University of Lisbon. The research interests are phenolic compounds from grapes and wines, wine technology and in particular the topic of wine aging process. Scientific Editor of two books of Oenology. He is author/co author of 40 articles in peer reviewer scientific journals, 10 book chapters and 65 communications in scientific congress. Member of editorial board of scientific journals. Supervisor of several Master Thesis in Oenology, Food Science and Food Technology. Develops consultancy services to wine companies and research projects evaluation.

Preface to "Phenolic Compounds in Fruit Beverages"

In this current special issue, different aspects related to phenolic compounds in fruit beverages are presented.

The phenolic compounds, which naturally occur in many fruit-based beverages, may affect positively or negatively the sensory characteristics of food, including fruit beverages, with important impacts on color, flavor, and astringency. Also, phenolic compounds are secondary metabolites abundant in our diet. An adequate consumption of phenolic compounds may offer health benefits. After the consumption of foods rich in phenolic compounds, such as fruit beverages, the colon is the main site of microbial fermentation. Phenolic compounds are transformed into phenolic acids or lactone structures by intestinal microbiota, which produce metabolites with biological and antioxidant activity, and evidence suggests those metabolites have health benefits for humans.

A large amount of different phenolic compounds, are responsible for physicochemical and sensory characteristics of fruit beverages. However, the phenolic composition of fruit beverages depends on several factors, namely, fruit cultivar, fruit maturity level, growing environment of fruits, cultural practices, postharvest conditions, and also processing and preservation techniques. Thus, the main goal of this special issue of *Beverages* is to present high-level research papers related to the phenolic composition of fruit beverages.

This special issue is composed of nine different works written by a group of international researchers in order to provide up-to-date reviews and current research on the different dimensions of phenolic compounds in fruit beverages. Thus, three interesting reviews are published by Canas [1], Ricci et al. [2], and Cosme et al. [3]. The first work discusses the influence of wood barrel characteristics, namely the botanical species of the wood and the toasting process that wood undergoes during the cooperage management. In particular, that work focuses on the impact of wood barrel composition on phenolic content of distilled wine spirits. The second review addresses the topic of the application of pulsed electric fields (PEF) on the red winemaking process, especially to improve the polyphenolic extraction and color release. Authors reported on data from several different PEF experiments in relation to grape varieties and process parameters, particularly on phenolic composition of red wines. Finally, the third review by Cosme et al. [3], focuses on grape juice phenolic composition, with a special focus on the potential beneficial effects on human health and on the grape juice sensory impact. In that work, several points are addressed, namely: grape juice production and phenolic composition, biological activity of phenolic compounds present in grape juices, and the impact of phenolic composition on sensory characteristics of grape juices.

The six remaining papers present current research in different dimensions of phenolic compounds from different fruit beverages. In this context, the work from Makebe et al. [4] analyzed optimization of the juice extraction process on must fermentation of overripe giant horn plantains by the use of an enzymatic extraction process and the impact on chemical composition of the wine produced, including total polyphenols and flavonoid content. In addition, the impact of the presence of phenolic compounds in the antioxidant activity of the produced wine also were studied. The second research paper from Zhang et al. [5] studied the application of poly-3,4-ethylenedioxythiophene (PEDOT) electrodes for the characterization of polyphenols in white wines. In that case, a Chardonnay white wine was used. According to these authors, the use of PEDOT electrodes showed considerable promise for applications in electroanalytical chemistry, based on their responsiveness to oxidizable substrates, such as wine polyphenols.

The antimicrobial and health-promoting activities of phenolic compounds are well known. Thus, Sabel et al. [6] studied the potential impact of a selection of phenolic compounds (which included hydroxybenzoic and cinnamic acids, stilbenes, and hydroxybenzaldehydes) on growth and viability of wine yeasts, and lactic and acetic acid bacteria. In addition, several factors that could determine the inhibitory effect of phenolic compounds—namely enzymatic oxidations, pH, and organic solvents—also were studied. The results obtained confirmed the antimicrobial activities of ferulic acid and resveratrol and showed the important role of syringaldehyde as an antimicrobial agent. The research work by Porto et al. [7] evaluated the physicochemical stability, antioxidant activity, and acceptance of a beet/orange juice during refrigerated storage. According to those authors, it was evident that the mixture of beet and orange juice contributed to an increase of juice stability, in particular as a result of high total phenolic compounds and antioxidant capacity.

Another work, by Stein-Chisholm et al. [8] analyzed the impact of rabbiteye blueberry juice processing on phenolic composition, in particular anthocyanin content. Those authors, using LC-MS/MS, identified and semi-quantified the major and minor anthocyanins at various steps of the juice production process. The results obtained showed that 10 major and 3 minor anthocyanins were identified in blueberry juice during the production process. In addition, it was also clear that increasing the number of steps in production induced a greater loss of anthocyanins in the final juice produced. For the last research work, Aguilar et al. [9] reported on sustainability. They characterized an antioxidant-enriched beverage made partly from grape musts and extracts of winery and grapevine by-products. Those authors try to find a novel functionality for grape pomace, grapevine leaves, and canes through their reuse as a functional matrix for the extraction of antioxidants. According to the results obtained, grape pomace and grapevine leaves are an important source of additional polyphenols in the preparation of enriched grape juice.

It was with great pleasure that I accepted the opportunity offered by MDPI, the publisher, to coordinate and serve as the guest editor of this special issue of *Beverages* regarding the theme of *Phenolic Compounds in Fruit Beverages*. I am greatly indebted to the authors who have generously shared their scientific knowledge and experience with others through their contribution to this special issue.

Conflicts of Interest: The authors declare no conflict of interest.

António Manuel Jordão

Special Issue Editor

References

1. Canas, S. Phenolic composition and related properties of aged wine spirits: Influence of barrel characteristics. A review. *Beverages* **2017**, *3*, 55. [CrossRef]
2. Ricci, A.; Parpinello, G.P.; Versari, A. Recent advances and applications of pulsed electric fields (PEF) to improve polyphenol extraction and color release during red winemaking. *Beverages* **2018**, *4*, 18. [CrossRef]
3. Cosme, F.; Pinto, T.; Vilela, A. Phenolic compounds and antioxidant activity in grape juices: A chemical and sensory view. *Beverages* **2018**, *4*, 22. [CrossRef]
4. Makebe, C.W.; Desobgo, Z.S.C.; Nso, E.J. Optimization of the juice extraction process and investigation on must fermentation of overripe giant horn plantains. *Beverages* **2017**, *3*, 19. [CrossRef]
5. Zhang, Q.; Türke, A.; Kilmartin, P. Electrochemistry of white wine polyphenols using PEDOT modified electrodes. *Beverages* **2017**, *3*, 28. [CrossRef]
6. Sabel, A.; Bredefeld, S.; Schlander, M.; Claus, H. Wine phenolic compounds: antimicrobial properties against yeasts, lactic acid and acetic acid bacteria. *Beverages* **2017**, *3*, 29. [CrossRef]
7. Porto, M.R.A.; Okina, V.S.; Pimentel, T.C.; Prudencio, S.H. Physicochemical stability, antioxidant activity, and acceptance of beet and orange mixed juice during refrigerated storage. *Beverages* **2017**, *3*, 36. [CrossRef]
8. Stein-Chisholm, R.E.; Beaulieu, J.C.; Grimm, C.C.; Lloyd, S.W. LC–MS/MS and UPLC–UV evaluation of anthocyanins and anthocyanidins during rabbiteye blueberry juice processing. *Beverages* **2017**, *3*, 56. [CrossRef]
9. Aguilar, T.; de Bruijn, J.; Loyola, C.; Bustamante, L.; Vergara, C.; Von Baer, D.; Mardones, C.; Serra, I. Characterization of an antioxidant-enriched beverage from grape musts and extracts of winery and grapevine by-products. *Beverages* **2018**, *4*, 4. [CrossRef]

beverages

MDPI

Review

Phenolic Composition and Related Properties of Aged Wine Spirits: Influence of Barrel Characteristics. A Review

Sara Canas [1,2]

[1] National Institute for Agrarian and Veterinary Research, INIAV-Dois Portos, Quinta da Almoínha, 2565-191 Dois Portos, Portugal; sara.canas@iniav.pt; Tel.: +351-261-712-106
[2] ICAAM—Institute of Mediterranean Agricultural and Environmental Sciences, University of Évora, Pólo da Mitra, Ap. 94, 7002-554 Évora, Portugal

Received: 20 September 2017; Accepted: 10 November 2017; Published: 14 November 2017

Abstract: The freshly distilled wine spirit has a high concentration of ethanol and many volatile compounds, but is devoid of phenolic compounds other than volatile phenols. Therefore, an ageing period in the wooden barrel is required to attain sensory fullness and high quality. During this process, several phenomena take place, namely the release of low molecular weight phenolic compounds and tannins from the wood into the wine spirit. Research conducted over the last decades shows that they play a decisive role on the physicochemical characteristics and relevant sensory properties of the beverage. Their contribution to the antioxidant activity has also been emphasized. Besides, some studies show the modulating effect of the ageing technology, involving different factors such as the barrel features (including the wood botanical species, those imparted by the cooperage technology, and the barrel size), the cellar conditions, and the operations performed, on the phenolic composition and related properties of the aged wine spirit. This review aims to summarize the main findings on this topic, taking into account two featured barrel characteristics—the botanical species of the wood and the toasting level.

Keywords: wine spirit; ageing; wooden barrels; oak wood; chestnut wood; toasting level; phenolic composition; chromatic characteristics; sensory properties; antioxidant activity

1. Introduction

The aged wine spirit is one of the most representative alcoholic beverages, taking into account production, trade [1], and consumption [2] worldwide. Its manufacture has a long history and a relevant socioeconomic role in the traditional wine countries, mainly in Europe. Among them, it is worth mentioning France and its regions of Armagnac and Cognac that date back to the 15th and 16th centuries, respectively [3–5], producing the most prestigious and top-selling aged wine spirits. In this scenario, Portugal and its Lourinhã region should also be highlighted, whose historical references on wine spirit production date back to the early 20th century; it was delimited in 1992 as an exclusive denomination for aged wine spirits, like the above-mentioned French regions [6].

According to the European legislation [7], the wine spirit can be aged for at least one year in wood containers or for at least six months in wood containers with a capacity of less than 1000 L. For wine spirits with geographical denomination, the ageing period is at least one year for Armagnac [8], and two years for Cognac [9] and Lourinhã [10].

Actually, the freshly distilled wine spirit is characterised by a high concentration of ethanol and richness of volatile compounds, but is devoid of phenolic compounds other than volatile phenols [11]. Ageing in a wooden barrel (of oak, chestnut, . . .) is traditionally included in wine spirit production technology, being recognized as a crucial step for adding value to the product. During this process, the beverage undergoes important modifications and becomes a complex mixture of hundreds of compounds in an ethanol-water matrix [12], leading to sensory fullness and improvement of its quality. There is much to know about the chemistry underpinning the ageing of wine spirits, but the scientific community unquestionably accepts that those physicochemical and sensory changes result from several phenomena [13–21] such as:

- Direct extraction of wood constituents;
- Decomposition of wood biopolymers (lignin, hemicelluloses and cellulose) followed by the release of derived compounds into the distillate;
- Chemical reactions involving only the wood extractable compounds;
- Chemical reactions involving only the distillate compounds;
- Chemical reactions between the wood extractable compounds and the distillate compounds;
- Evaporation of volatile compounds and concentration of volatile and non-volatile compounds;
- Formation of a hydrogen-bonded network between ethanol and water.

Among these phenomena, the release of wood extractable compounds into the wine spirit, namely low molecular weight phenolic compounds and tannins, plays a decisive role in its chemical composition, sensory properties [22–24] and overall quality. In addition, oxidation reactions involving these compounds and those of the distillate are of paramount importance [25–29]. They are triggered by the slow and continuous diffusion of oxygen through the space between staves and through the wood [30–32].

The research carried out over the last decades has shown that the aforementioned changes are closely related to the action of factors ruling the ageing process, namely:

(i) The wooden barrel characteristics—the wood botanical species used, and the characteristics imparted by the cooperage technology (especially the seasoning/maturation of the wood and the heat treatment of the barrel), and the barrel size [33];

(ii) The cellar conditions—temperature, relative humidity and air circulation [18,34,35];

(iii) The technological operations performed during the ageing period, such as the refilling with the same wine distillate to offset the loss by evaporation [18,20,36], the addition of water to decrease the alcoholic strength [37], and stirring to homogenize the wine spirit and to enhance the extraction of wood compounds [38].

Concerning the resulting sensory properties, positive correlations between the phenolic composition and the color were established [23,39,40], which are in accordance with the findings in studies on Porto wine [41] and wine [42–46]. Notwithstanding the intricate effect of compounds on aroma and flavour owing to the complexity of their interactions and the multiple sensations involved [47,48], some features have been often related to the phenolic composition of the aged wine spirit and of other aged beverages. There is evidence on the relationship between: the vanilla aroma and vanillin concentration [24,49,50]; the bitterness and phenolic acids, their ethyl esters, and (+)-lyoniresinol concentrations [48,51–54]. Among these properties, the vanilla aroma should be highlighted due to its outstanding importance for aged wine spirit quality [55–57]. The relationship between astringency and ellagitannin and gallotannin concentrations is still unclear for thesekind of beverages [48,51–54,58,59].

In addition, the phenolic compounds (in this case almost exclusively extracted from the wood) exhibit a wide range of biological effects, many of which have been ascribed to their antioxidant activity. Several studies mention the antioxidant activity of some phenolic acids [60–69], phenolic aldehydes [70,71], coumarins [72], tannins [65,67,73–75], lignans [76], and of some volatile phenols [77]. This topic is of great relevance in a spirit drink, since the harmful effect of high alcoholic strength on consumer's health can be offset by the intake of such bioactive compounds [74,78–84]. Such benefits are expected in beverages that have undergone ageing in wood, especially wine spirits and whisky, but not in the traditional gins and vodkas [78].

Despite the knowledge acquired through different studies on the phenolic composition and related properties of the aged wine spirit modulated by the ageing technology, so far there have been no published articles systematizing it. This review assembles the main findings on the topic, taking into account two featured barrel characteristics—the wood botanical species and the toasting level.

2. Phenolic Compounds Found in Aged Wine Spirits

Several low molecular weight phenolics (Table 1) have been identified and quantified in aged wine spirits using high performance liquid chromatography (HPLC) or capillary electrophoresis (EC). All of them are non-flavonoid compounds; therefore, the phenolic composition of the aged wine spirit differs from that of wine and wine aged in wood, in which flavonoid and non-flavonoid compounds coexist [85,86].

Table 1. Low molecular weight phenolic compounds found in aged wine spirits.

Class	Compound	Concentration Range *	References	Wine Spirits **
Phenolic aldehydes	Sinapaldehyde	0.05–42.31	[13,15,29,38,74,87–105]	a,b,c,d,e,j,l
	Syringaldehyde	0.20–34.20	[13,15,17,29,38,74,87–94,98–105]	a,b,c,d,e,j,l
	Vanillin	0.10–18.40	[13,15,17,29,38,87–94,97–105]	a,b,c,d,e,j,k,l
	Coniferaldehyde	0.05–12.94	[13,15,29,38,74,87–89,91–93,98–100,102,103,105]	a,b,c,d,e,j,l
Phenolic acids	Gallic acid	1.00–168.67	[15,17,38,74,96,97,99,101,102,105]	a,c,f,k,l
	Ellagic acid	3.90–104.00	[38,74,97,99,101,102,105]	a,c,k,l
	Syringic acid	0.40–17.18	[15,17,29,38,74,88,89,99–102,105]	a,b,c,d,l
	Vanillic acid	0.20–10.95	[15,17,29,38,74,88,89,99–102,105]	a,b,c,d,l
	Ferulic acid	0.05–9.94	[15,88,89,102,105]	a,b,c
	Protocatechuic acid	0.12–2.27	[15]	a
	Coumaric acid	0.02–1.20	[15,88]	a,b
Coumarins	Scopoletin	6.00–301.10	[38,74,93–95,99,102]	a,b,c,e,f,g,h,i,l
	Umbelliferone	0.11–7.00	[38,93,102]	a,b,c
Lignans	Lyoniresinol	3.40–17.50	[92,99,105]	c,j,l
Phenyl ketones	Acetovanillone	0.51–6.21	[107]	a

* Concentration in mg/L except for coumarins, which are in µg/L; the compounds are arranged in descending order of quantitative importance within each class; ** a—Cognac; b—Armagnac; c—Lourinhã; d—Moldovan; e—American; f—Spanish; g—Bulgarian; h—Canadian; i—Russian; j—Japanese; k—French; l—wine brandy.

Regardless the ageing conditions and the analytical methodologies employed, the results presented in Table 1 show that phenolic acids are the most abundant phenolic compounds in wine spirits, accounting for ca. 70% of low molecular weight phenolic compounds, followed by phenolic aldehydes (ca. 15%), lignans (ca. 12%), phenyl ketones (ca. 3%) and coumarins (0.1%) (Figure 1).

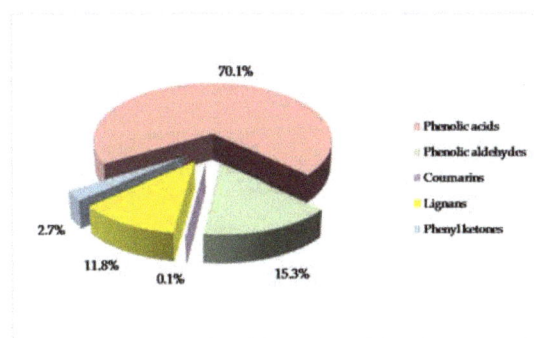

Figure 1. Relative importance of some phenolic classes in the aged wine spirits.

These acids, aldehydes and coumarins already exist in oak and chestnut heartwoods in the free form or linked to parietal constituents [108]. Nevertheless, their contents change considerably through the thermal degradation of the wood lignin [109–113] together with the increase of wood permeability [114,115] during the heat treatment of the barrel. Hence, higher amounts can be released into the wine spirit over the ageing process. Besides, lignin's hydrolysis occurring during ageing may also contribute to the enrichment in some phenolic aldehydes and phenolic acids [14,29,116].

Gallic acid and ellagic acid are the most representative phenolic acids, followed by syringic, vanillic, and ferulic acids. Protocatechuic acid and coumaric acid seem to be less important since they have not always been detected in the aged wine spirit.

The syringyl-type aldehydes (sinapaldehyde and syringaldehyde) are more plentiful than the guaiacyl-type aldehydes (vanillin and coniferaldehyde), probably due to the higher thermal stability of the former and their consequently greater accumulation in the toasted wood [117,118].

Among coumarins, the literature indicates scopoletin as more abundant than umbelliferone [38,93,102].

Concerning the lignans, few works report the presence of lyoniresinol in aged wine spirits. Despite the non-negligible amounts found, more advanced analytical conditions required for the separation and quantification of its two enantiomers [53,107,119] may justify their non-detection by HPLC under common chromatographic conditions. One of the enantiomers, (+)-lyoniresinol, was also quantified in oak wood [53,92,107], in which it remains stable under toasting until 200 °C [107,117].

As far as we know, acetovanillone was only quantified in small amounts in one study [106] and was also detected in toasted wood [14,120], being mainly formed by the thermal degradation of lignin [14,121].

In addition to the low molecular weight phenolic compounds, five hydrolysable tannins were identified and quantified in aged wine spirits by liquid chromatography coupled to mass spectrometry (LC-MS) and HPLC, respectively [122] (Table 2).

Table 2. Hydrolysable tannins found in aged wine spirits.

Class	Compound	Concentration Range
Ellagitannins	Castalagin	2.81–20.75
	Vescalagin	0.03–0.24
	Roburin E	0.08–0.19
	Grandinin	0.06–0.16
Gallotannins	Monogalloyl-glucose	0.47–5.95

Concentration in mg/L gallic acid. Adapted from [122].

The four monomeric ellagitannins and the monomeric gallotannin are derived from the wood (oak or chestnut), in which they are present in higher amounts, together with four dimeric ellagitannins (roburins A, B, C and D) [67,123–126], and with other dimeric and trimeric gallotannins [127,128]. Therefore, their low content or absence in the aged wine spirits may result from the thermal degradation during the heat treatment of the wood [117,126,129–131]. Moreover, low extraction [132], as well as oxidation and hydrolysis of ellagitannins, may occur during ageing [99,101].

Among the identified ellagitannins, castalagin and vescalagin are the most representative ones, followed by xylose and lyxose derivatives (roburin E and grandinin, respectively), as in the wood. Taking into account the total average content of soluble ellagitannins in Cognacs (4–840 mg/L; [15,101]) and in Armagnacs (155–702 mg/L; [22]), the quantified ellagitannins (Table 2) only represent ca. 3%, 0.03%, 0.03% and 0.03%, respectively; that is, other unidentified ellagitannins are likely to be extracted from the wood into the wine spirit and to have a greater contribution to the total amount. Actually, so far, few soluble tannins have been found in the aged wine spirit, contrasting with the increasing number of tannins and derived compounds detected in the aged wine [43,133].

Data from Tables 1 and 2 clearly illustrate the huge variability of the phenolic compounds concentrations in aged wine spirits. Indeed, it encompasses the effect of the analytical methodology used, but especially the impact of the barrel characteristics. Details on the last aspect and its repercussion on the chromatic characteristics, color, other related sensory properties, and antioxidant activity of the aged wine spirit are presented in the following sections.

3. Influence of the Wood Botanical Species

The wood most commonly used for the ageing of wine spirits is from the oak species *Quercus robur* L., principally from the French region of Limousin [37,134]. However, other wood botanical species have been increasingly studied to evaluate their potential for the cooperage, focusing their chemical composition: *Quercus sessiliflora* Salisb., particularly from the French region of Allier, and *Quercus alba* L., mainly from North America [67,108,135,136]; *Quercus pyrenaica* Willd., grown in Mediterranean countries [67,108,124,126,137]. Chestnut wood (*Castanea sativa* Mill.) has also been exploited for this purpose, and is of particular significance in the countries bordering the Mediterranean Sea due to historical, economical, and social aspects of its cultivation [138]. Its suitability for the cooperage aiming the ageing of wine spirit has also been investigated [6,102,108,120,139,140].

3.1. Phenolic Composition

Considerable attention has been devoted by several research teams to the chemical composition of wood used in oenology. However, only a limited number of studies about its impact on the chemical composition of the aged wine spirit have been published. Moreover, their experimental designs are not always fully described, hindering the comparison of results obtained in different approaches. To the best of our knowledge, the exception lies in four older works that examined the effect of one kind of wood. Baldwin et al. [13] found low levels of vanillin (ranging from 0.6 to 1.5 mg/L), syringaldehyde (varying from 1.2 to 7.6 mg/L), coniferaldehyde (ranging from 0.3 to 1.8 mg/L), and sinapaldehyde (ranging from 0.2 to 3.4 mg/L) in American wine spirits aged in new barrels of American oak. Similarly, Nabeta et al. [92] reported low average contents of vanillin (0.6 mg/L), syringaldehyde (0.35 mg/L), coniferaldehyde (1.4 mg/L), and sinapaldehyde (0.8 mg/L) in Japanese wine spirits aged over a six-year period in new barrels of French oak wood. Tricard et al. [94] also observed low amounts of vanillin (2.28 mg/L) and syringaldehyde (6.60 mg/L), but a considerable amount of scopoletin (109 mg/L), in Cognacs aged over a seven-year period in new barrels of French oak. In contrast, Puech and Moutounet [99] found higher contents of vanillin (8.2 mg/L), syringaldehyde (19.4 mg/L), coniferaldehyde (17.8 mg/L), and sinapaldehyde (19.8 mg/L) in wine brandies aged over a seven-year period in new barrels of Limousin oak. They also found high contents of ellagic acid (62.9 mg/L), gallic acid (31.0 mg/L), vanillic acid (7.9 mg/L), and syringic acid (8.2 mg/L).

The most recent works [6,56,105] were based on a factorial design using the same wine distillate from the Lourinhã region (produced by Adega Cooperativa da Lourinhã) aged over a four-year period in barrels made from the following kinds of wood: Limousin oak (*Q. robur* L.) and Allier oak (*Q. sessiliflora* Salisb.) from French forests; American oak (mixture of *Q. alba* L./*Q. Stellata* Wangenh. and *Q. lyrata* Walt./*Q. bicolor* Willd.) from Pennsylvania/USA; Portuguese oak (*Q.pyrenaica* Willd.) and chestnut (*C. sativa* Mill.) from the North of Portugal. The 250 L barrels were supplied by J. M. Gonçalves cooperage (Palaçoulo, Portugal) and were placed in the cellar of Adega Cooperativa da Lourinhã in similar environmental conditions.

Significant differences in the contents of the majority of low molecular weight phenolic compounds of the aged wine spirit according to the wood used were observed (Table 3). Chestnut wood induced the highest content of phenolic acids in the wine spirit, especially of gallic acid and ellagic acid, as noticed for the ageing of red wine [141,142]. The highest levels of vanillin and syringaldehyde were also found in the wine spirit aged in chestnut barrels. Portuguese oak wood promoted intermediate enrichment, with the highest level of sinapaldehyde, while the other kinds of oak had a weaker performance. However, the richness of coumarins associated with the American oak should be stressed, because these compounds can act as chemical markers of this kind of wood [56,95].

Table 3. Mean concentrations of low molecular weight compounds in wine spirits aged four years in different kinds of wood.

Compound	American oak	Allier Oak	Limousin Oak	Portuguese Oak	Chestnut
Ellagic acid	32.45 [a]	37.19 [a]	49.38 [b]	81.16 [c]	91.27 [d]
Gallic acid	10.49 [a]	13.46 [a]	11.52 [a]	37.80 [b]	218.19 [c]
Vanillic acid	2.97 [b]	2.04 [a]	2.62 [a,b]	2.96 [b]	6.15 [c]
Syringic acid	3.56 [a]	3.58 [a,b]	4.09 [a,b,c]	5.03 [c]	19.77 [d]
Ferulic acid	2.97 [a]	2.85 [a]	3.03 [a]	6.06 [b]	6.39 [c]
Vanillin	6.21 [a,b]	5.50 [a]	6.33 [b]	6.41 [b]	8.28 [c]
Syringaldehyde	15.06 [b]	11.73 [a]	15.02 [b]	14.94 [b]	15.89 [b]
Coniferaldehyde	9.04	8.27	9.12	8.75	7.78
Sinapaldehyde	16.71 [b]	14.63 [a,b]	16.76 [b]	19.65 [c]	11.94 [a]
Umbelliferone	1.48 [c]	0.78 [a]	0.95 [b]	0.98 [b]	0.92 [b]
Scopoletin	164.77 [d]	19.74 [b]	37.12 [c]	10.33 [a]	8.63 [a]
ΣLMW	122.68 [a]	127.68 [a]	144.03 [a]	224.0 [b]	395.93 [c]

Concentration in mg/L absolute ethanol except for coumarins, which are in µg/L absolute ethanol; mean values (n = 24) followed by different letters ([a], [b], [c], [d]) in a row are significantly different ($p < 0.05$); ΣLMW—Sum of low molecular weight phenolic compounds concentrations. Adapted from [56].

Comparing the results of the oldest and most recent works, important differences in the concentrations of phenolic aldehydes, phenolic acids and scopoletin in wine brandies aged in French oak wood and American oak wood are observed. They may express the effect of the geographical origin of the wood, as well as the interaction between the kind of wine distillate and the wood, in the extraction kinetics of such compounds. Despite the observed variability, the American oak wood had a lesser contribution to the phenolic composition of the aged wine spirit. Taking into account the results of Puech and Moutounet [99], the performance of Limousin oak wood resembled that of Portuguese oak wood.

The phenolic differentiation of wine spirits was ascribed to the pool of phenolic compounds in the different kinds of wood under study [108,126,140,143] and to lignin hydrolysis during ageing [14,29,102,116,144]. Furthermore, gallic acid and ellagic acid can be directly extracted from the wood, or derived from the hydrolysis of gallotannins [112] and ellagitannins [101], respectively, especially in the first years of ageing [22,101].

Comparing the hydrolysable tannins of the wine spirit aged in Limousin oak wood and in chestnut wood, Canas et al. [122] did not observe significant differences in their contents (Table 4). Monogalloyl-glucose was only present in the wine spirit aged in chestnut barrels, as in the corresponding wood [123], although detection of this monomeric gallotannin in *Quercus robur* wood already made by other authors [127]. The robustness of the methods used in the isolation and quantification of the tannin fraction pointed to the conclusion that the above-mentioned effect could be caused by the wood intraspecific variability [145].

Table 4. Mean concentrations of hydrolysable tannins in wine spirits aged four years in different kinds of wood.

Compound	Limousin Oak	Chestnut
Castalagin	12.07	6.33
Vescalagin	0.11	0.17
Roburin E	0.14	0.12
Grandinin	0.12	0.14
Monogalloyl-glucose	nd	5.16

Concentration in mg/L gallic acid; mean values (n = 9); nd—not detected. Adapted from [122].

3.2. Chromatic Characteristics and Sensory Properties

The color is a core element in the perception of wine spirit quality, as for other beverages [146]. It determines the first impression and rules the consumer's choice [147]. Among the few authors who studied it in aged wine spirits [39,40,148], Canas et al. [23] remarked that the chromatic characteristics (CIELab parameters) are closely related to the kind of wood used (Figure 2). In this investigation [23], the wine spirit aged in chestnut wood stood out by its more evolved color than the wine spirits aged in oak wood. It displayed higher color intensity (lower L*), higher red (a*) and yellow (b*) hues, and higher saturation (C*) that made it look older than the latter. Indeed, there is scientific evidence about the color evolution of wine spirits over the ageing time, which is marked by a decreasing of lightness and an increasing of saturation, red hue and yellow hue [148]. Among the wine spirits aged in oak wood, the one aged in Portuguese oak exhibited greater evolution of the chromatic characteristics than those aged in Limousin oak, American oak and Allier oak. Once the wine distillate itself is colorless due to the absence of phenolic compounds other than volatile phenols [11], the observed changes are assigned to the wood stage. Recently, Rodríguez-Solana et al. [149] reported concordant results for a grape marc spirit aged in *Q. robur*, *Q. alba* and *Q. petraea* wooden barrels.

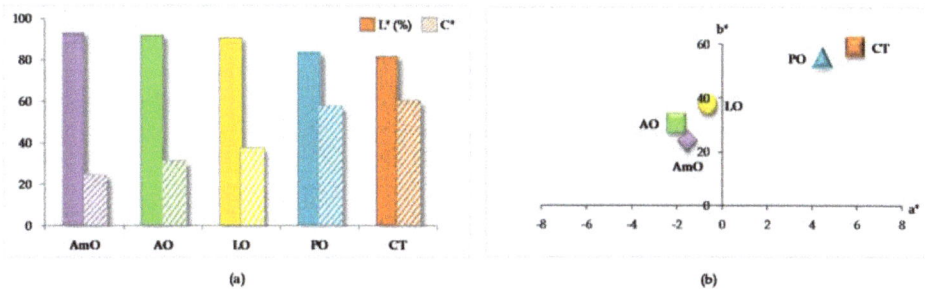

Figure 2. Mean values (*n* = 24) of chromatic characteristics of the wine brandies aged in different kinds of wood: (**a**) lightness (L*) and saturation (C*); (**b**) chromaticity coordinates (a*, b*); AmO—American oak; AO—Allier oak; LO—Limousin oak; PO—Portuguese oak; CT—Chestnut. For each chromatic characteristic, the differences between the aged wine spirits are very significant (*p* < 0.01). Adapted from [23].

The different pool of phenolic compounds and extraction kinetics (Table 3), the oxidative phenomena occurring during ageing [26,27,101], and the condensation reactions between phenolic compounds promoted by each kind of wood possibly accounted for the differences in the chromatic characteristics. Condensation reactions between tannins mediated by acetaldehyde (resulting from ethanol oxidation and which may represent more than 90% of the aldehydes' content of the aged wine spirit [14,16]), by phenolic aldehydes and furanic derivatives, such as furfural and 5-hydroxymethylfurfural, are likely to happen, as in wine during ageing [146,150–152].

Interestingly, as described in the same work [23], the color perceived by the tasters (Figure 3) was consistent with the chromatic characteristics of the aged wine spirits (Figure 2).

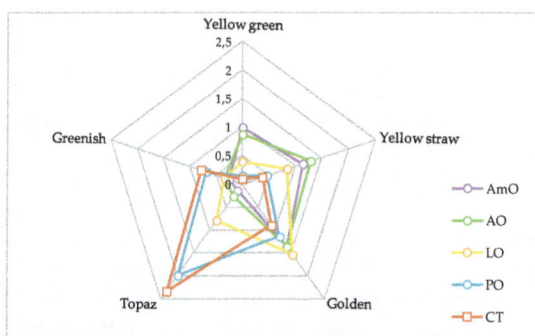

Figure 3. Color profile based on the average panel score's attributes of the wine spirits aged four years in different kinds of wood. For each color attribute, the differences between the aged wine spirits are very significant ($p < 0.01$). Adapted from [23].

The wine spirits aged in chestnut barrels were characterized by a more evolved color owing to the highest intensities of topaz (orange, amber color), greenish color, and the lowest intensities of golden and yellow straw (Figure 3). The wine spirit aged in Portuguese oak wood had high intensity of topaz and golden while those aged in the other kinds of oak wood showed higher intensities of golden, yellow straw and yellow green. Topaz is the main color of older wine spirits, resulting from the combination of higher positive values of chromaticity coordinates a* (red hue) and b* (yellow hue), whereas the golden prevails in wine spirits with less ageing time. Therefore, as for the analytical color, the sensory color of the wine spirits expressed the differences in their phenolic composition.

Examining other sensory properties of the wine spirits, Caldeira et al. [153] found significantly higher intensities of vanilla and astringency underlying the cluster formed by the wine spirits aged in Portuguese oak and chestnut. For these attributes, the wine spirits aged in Allier oak and American oak presented the opposite features, while those aged in Limousin oak showed an intermediate profile. Similar results were obtained for grape marc spirits aged in *Q. robur*, *Q. alba* and *Q. petraea* barrels [149]. The observed behavior for the vanilla aroma is ascribed to the vanillin content (Table 2), while the astringency seems not be related to the content of hydrolysable tannins (Table 3).

3.3. Antioxidant Activity

The antioxidant activity of some spirit drinks, namely the aged wine spirit, has aroused the interest of the researchers. Indeed, the "French Paradox" showed that for apparently the same level of risk factors, cardiovascular mortality rate is lower in France than in the European Northern countries [154]. Moreover, among the French regions, this phenomenon was particularly evident in the southwest France, a region where people do not drink more wine than elsewhere, but often drink Armagnac [155]. Therefore, some research was conducted to evaluate the antioxidant activity of Armagnacs and Cognacs [74,78,80] as well as of Spanish brandies [82,156], revealing its positive correlation with the ageing time. Later on, the influence of the wood botanical species used was examined [122]. Similar studies were also done with brandies [84,157] but the corresponding experimental designs raise many doubts, and do not allow comparison of outcomes with those obtained for the aged wine spirits.

Studying the in vitro antioxidant activity of the same wine spirit aged in chestnut barrels and in Limousin oak barrels, Canas et al. [122] concluded that the wood botanical species induced significantly different antioxidant activity (measured by 1,1-diphenyl-2-picrylhydrazyl radical scavenging activity—DPPH) regardless the toasting level. The antioxidant activity promoted by the chestnut wood (DPPH inhibition = 93.5%) was two-fold higher than that promoted by Limousin oak wood (DPPH inhibition = 45.7%)—Figure 4.

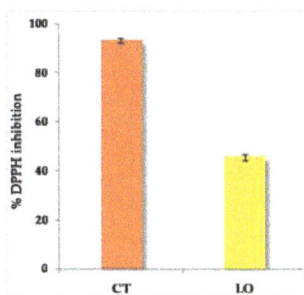

Figure 4. Mean values of antioxidant activity of the wine spirits aged in different kinds of wood (LO—Limousin oak; CT—chestnut). Adapted from [122].

This effect is explained by the highest phenolic content of the wine spirit aged in chestnut barrels; synergistic phenomena [81,158] and antagonistic phenomena [159] between individual phenols are also likely to occur. The antioxidant activity was mainly correlated with gallic acid ($r = 0.9555$) and ellagic acid ($r = 0.6138$), whose contents were significantly higher in the wine spirit aged in chestnut wood (Table 3). Actually, these phenolic acids are well-known bioactive compounds [61,62,66,69,160]. Syringaldehyde may also have contributed to this feature [70,71], especially in the wine spirit aged in chestnut barrels. It is interesting to note that a similar relationship of the antioxidant capacity was reported by Rodríguez Madrera et al. [143] for *C. sativa* and *Q. robur* wood extracts based on the levels of phenolic acids and phenolic aldehydes. Regarding the role of hydrolysable tannins in the observed behavior, no significant correlations were pointed out by Canas et al. [122] for individual compounds and for the total content. However, strong correlations between the antioxidant capacity and monomeric ellagitannins (castalagin, vescalagin, roburin E and grandinin), and dimeric ellagitannins (roburins A–D) were emphasized for wood extracts including *C. sativa* and *Q. robur* [67]. On the other hand, Da Porto et al. [74] stated that ellagitannins are the major contributors to the overall antioxidant activity of Cognac. So, this kind of discrepancy could be justified by the variability of wood composition [145] plus the variability induced by the wood heat treatment together with differential extraction from the wood and reactions involving ellagitannins during ageing, as aforementioned.

4. Influence of the Heat Treatment of the Barrel

Research on oak wood and chestnut wood have shown that the heat treatment of the barrel is of remarkable importance to the pool of extractable compounds that can be released into the beverage when it contacts the wood [102,109–113,117,118].

The heat treatment is part of the barrel making process, being performed by the French technique using fire, or by the American technique using heated steam for bending the staves followed by fire [161]. In European cooperage, the barrel is heated over a fire of wood shavings with various techniques of spraying or swabbing with water to enable the bending of the staves to the concave shape of a barrel without breaking—the bending phase [110,161]. Then, the barrel is placed again over the fire to heat the inner surface and to cause significant toasting in order to modify the structure [114], the physical properties [115], and the chemical composition of the wood [109,110,121], which confer a distinct character to the wine or distillate aged in it—the toasting phase. Despite the diversity of toasting protocols, the toasting level is usually classified as light, medium or heavy. In practice, the result mostly depends on the binomial temperature/time applied to each wood botanical species [109,110,143,162,163].

4.1. Phenolic Composition

Scientific data about the influence of the wood toasting level on the phenolic composition and related properties of the aged wine spirit are rather scarce. Notwithstanding, older works made on Spanish brandies by Artajona et al. [96], on Cognacs by Cantagrel et al. [164] and Viriot et al. [101], and on French wine spirits by Rabier and Moutounet [97] and Puech et al. [165] are noteworthy. Artajona et al. [96] found increasing contents of phenolic aldehydes in brandies with an increasing of barrel toasting intensity: ca 18 mg/L, 30 mg/L, and 58 mg/L under the influence of light, medium and heavy toasting, respectively. Rabier and Moutounet [97] observed increasing contents of ellagic acid (ca 15 mg/L and 60 mg/L), gallic acid (ca 4 mg/L and 9 mg/L), and vanillin (ca 0.5 mg/L and 1 mg/L) in a wine spirit aged over a two-year period in new oak barrels with light and heavy toasting levels. Puech et al. [165] also studied the influence of the toasting level (light, medium and heavy) in a wine spirit aged over a two-year period in new barrels of Limousin oak. They found an increasing content of vanillin with the rise of toasting intensity, which remained below 5 mg/L; a similar behavior was observed for syringaldehyde with ca.1 mg/L, 7 mg/L and 11 mg/L under the effect of light, medium and heavy toasting, respectively; for coniferaldehyde and sinapaldehyde, a sharp increase between light and medium toasting (from ca 3 to ca 13 mg/L, and from ca 2 to ca 22 mg/L, respectively) and a slight decrease under heavy toasting (ca 11 mg/L and 21 mg/L, respectively) were described.

In recent years, a comprehensive investigation was performed [6,56,120]. In that study, the same wine distillate from *Lourinhã* region (produced by Adega Cooperativa da Lourinhã) was aged over a four-year period in 250 L barrels. The barrels were made by J. M. Gonçalves cooperage (Palaçoulo, Portugal) using the following kinds of wood: Limousin oak (*Q. robur* L.) and Allier oak (*Q. sessiliflora* Salisb.) from French forests; American oak (mixture of *Q. alba* L./*Q. Stellata* Wangenh. and *Q. lyrata* Walt./*Q. bicolor* Willd.) from Pennsylvania/USA; Portuguese oak (*Q. pyrenaica* Willd.) and chestnut (*C. sativa* Mill.) from the North of Portugal. These barrels were divided into three groups. Then, each group were submitted to one of the three levels of toasting—light (LT), medium (MT) and heavy (HT)—according to the cooperage protocol: 10 min for light toasting, 20 min for medium toasting and 25 min for heavy toasting [120]. They were filled with the same wine distillate and kept in the cellar of Adega Cooperativa da Lourinhã in similar environmental conditions.

Analysing the low molecular weight phenolic compounds of the wine spirits aged in them, Canas [56] showed that the toasting level had a significant effect on the concentration of all phenolic compounds, except for scopoletin (Table 5), confirming the results of previous studies [96,97,164,165].

Table 5. Mean concentrations of low molecular weight compounds in wine spirits aged four years in barrels with different toasting levels.

Compound	Light Toasting	Medium Toasting	Heavy Tosating
Ellagic acid	38.85 [a]	57.69 [b]	87.36 [c]
Gallic acid	42.27 [a]	55.44 [b]	54.22 [b]
Vanillic acid	2.05 [a]	3.20 [b]	4.48 [c]
Syringic acid	5.08 [a]	6.00 [b]	8.64 [c]
Ferulic acid	4.30 [a]	4.49 [a]	5.00 [b]
Vanillin	2.94 [a]	6.43 [b]	10.07 [c]
Syringaldehyde	4.31 [a]	13.00 [b]	26.28 [c]
Coniferaldehyde	3.25 [a]	8.64 [b]	13.55 [c]
Sinapaldehyde	3.99 [a]	14.49 [b]	30.60 [c]
Umbelliferone	0.47 [a]	0.92 [b]	1.64 [c]
Scopoletin	37.36	39.42	36.58
ΣLMW	117.82 [a]	197.10 [b]	295.08 [c]

Concentration in mg/L absolute ethanol except for coumarins, which are in µg/L absolute ethanol; mean values (*n* = 56) followed by different letters ([a], [b], [c]) in a row are significantly different (*p* < 0.05); ΣLMW—Sum of low molecular weight phenolic compounds concentrations. Adapted from [56].

Furthermore, Viriot et al. [101] and Canas [56] emphasized a positive relationship between ellagic acid concentration in the wine spirits and the toasting intensity of the barrel. A different pattern is identified for gallic acid; its concentration in the aged wine spirits increases under the influence of medium toasting and slightly decreases under heavy toasting. Recent results obtained for grape marc spirit corroborate it [149]. This pattern expresses the behavior of gallic acid in the toasted wood, in which it undergoes degradation from the medium toasting as a consequence of higher thermal sensitivity [97,166]. In contrast, higher level of ellagic acid is ascribed to its high fusion point and greater accumulation in the toasted wood, as observed by Rabier and Moutounet [97]. As in the untoasted wood [108] and toasted wood [109,137], ellagic acid and gallic acid still remain the major phenolic acids of the aged wine spirits, being mainly derived from the wood ellagitannins and gallotannins [118,129,130,167].

It was also demonstrated that the rise of toasting level of the barrel promoted an increase of vanillic acid, syringic acid, ferulic acid and phenolic aldehydes contents in the aged wine spirits, as in the aforementioned studies, except for coniferaldehyde and sinapaldehyde [165]. It is well-known these compounds resulted from the wood lignin's decomposition [117] (Figure 5). Under mild temperatures, decarboxylation and cleavage of the aryl-alkyl ether bonds of the terminal units of this biopolymer take place, originating the cinnamic aldehydes (coniferaldehyde and sinapaldehyde). At higher temperatures, an oxidative cleavage of double C-C bond of the aliphatic chain of these aldehydes may occur, yielding the corresponding benzoic aldehydes (vanillin and syringaldehyde). The resulting concentrations express the balance between synthesis and degradation reactions. Therefore, the slight decrease of coniferaldehyde and sinapaldehyde contents reported by Puech et al. [165] for heavy toasting should have resulted from specificity of the toasting protocol, which induced higher degradation of these aldehydes. As the temperature rises, the phenolic aldehydes thus formed give rise, by decarboxylation, to the corresponding phenolic acids. Hence, they accumulate in the toasted wood [109,130,137,163].

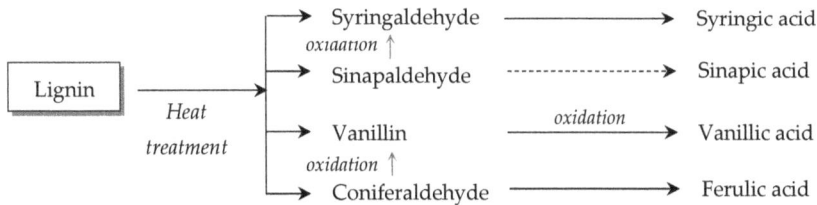

Figure 5. Mechanism of lignin's decomposition and formation of derived compounds; proposed by [117].

Furthermore, higher permeability of the wood and better access of the wine spirit to wood extraction sites caused by fragmentation of cell structures and reorganization of lignocellulose network [114,115] may also facilitate their release into the wine spirit. Likewise, lignin's hydrolysis during the ageing period may contribute to their increase in the beverage [14,29,102,116,144]. The presence of oxygen and the mild acidity of the medium, mainly modulated by the increase of acetic acid content over time, favor this pathway [37].

Regardless the toasting level of the barrel and the ageing time, it has been found [29,56,102] that syringyl-type aldehydes (sinapaldehyde and syringaldehyde) prevailed over those of guaiacyl-type (vanillin and coniferaldehyde) in the aged wine spirits. On the other hand, an increase in the syringyl/guaiacyl ratio with the toasting intensity has been referred [14,56,97]. In the above-mentioned work [56], mean values of 1.34, 1.82 and 2.41 were obtained for the same wine spirit aged during four years in barrels with light, medium and heavy toasting levels, respectively. This suggests that higher thermal stability of the syringyl compounds and subsequent higher availability in the toasted wood [117,118] was the causal effect.

There is also evidence of the increase of umbelliferone content in the wine spirit with the toasting level of the barrel [56], but the chemical mechanisms underpinning its formation/degradation during the heat treatment of the wood are still unknown.

Concerning the hydrolysable tannins, no significant differences in wine spirits aged in barrels with different toasting level were reported [122].

4.2. Chromatic Characteristics and Sensory Properties

As for the wood botanical species, data from the literature [23] show the modulating effect exerted by the toasting level of the barrel on the chromatic characteristics of this beverage (Figure 6). The higher the toasting level of the barrel the more the evolution of the aged wine spirit color (higher intensity, saturation and red and yellow hues) with significant increments between levels. Acquisition of these chromatic characteristics makes the wine spirit aged in heavy toasting barrels look older than those aged in medium and light toasting barrels (as noticed for the wine spirit aged in chestnut wood when compared with those aged in different kinds of oak wood). These outcomes are correlated with the phenolic compounds extracted from the wood (Table 5) and are in agreement with those obtained for wine aged in barrels with different toasting levels [167]. In addition, the oxidative phenomena underlying the ageing process may also be responsible for the color acquired by the aged wine spirit; the higher the toasting level the higher the wood permeability to oxygen [114,115], and therefore greater extension of oxidation reactions are expected.

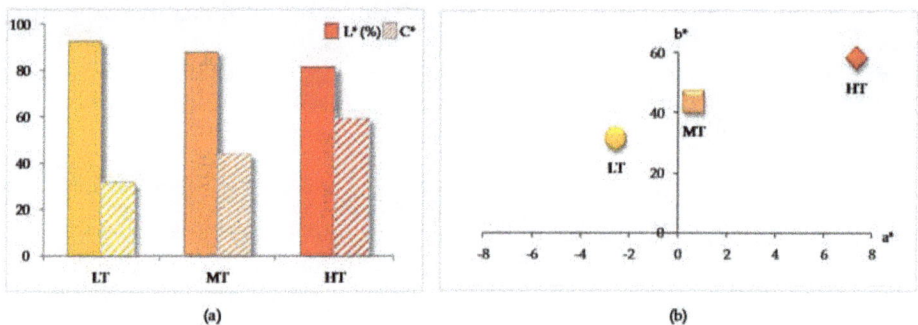

Figure 6. Mean values (*n* = 56) of chromatic characteristics of the wine spirits aged in barrels with different toasting levels: (**a**) lightness (L*) and saturation (C*); (**b**) chromaticity coordinates (a*, b*); LT—light toasting; MT—medium toasting; HT–heavy toasting. For each chromatic characteristic, the differences between the aged wine spirits are very significant ($p < 0.01$). Adapted from [23].

From the sensory point of view, Canas et al. [23] indicated the predominance of yellow straw and yellow green in the wine spirits aged in light toasting barrels, and the prevalence of golden and topaz in those aged in medium and heavy toasting barrels, respectively (Figure 7). These results are consistent with those obtained by the CIELab method (Figure 6), showing a faster ageing of the wine spirit associated with the heavy and medium toasting levels.

Other sensory properties related to the phenolic composition, such as the vanilla aroma and astringency, had higher intensities associated with the heavy toasting barrels [153]. The wine spirits aged in light toasting barrels and medium toasting barrels revealed opposite and intermediate intensities of these attributes, respectively. Increasing concentrations of vanillin with the toasting intensity (Table 5) should explain the effect on the vanilla aroma. Regarding astringency, the existing information does not allow the establishment of a reliable relationship with the phenolic composition.

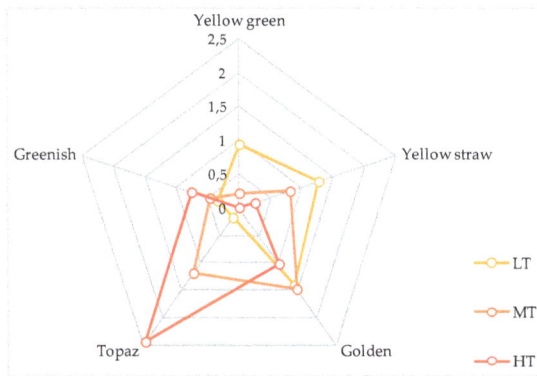

Figure 7. Color profile based on the average panel score's attributes of the wine spirits aged during four years in barrels with different toasting levels (LT—light toasting; MT—medium toasting; HT—heavy toasting). For each color attribute, the differences between the aged wine spirits are very significant ($p < 0.01$). Adapted from [23].

4.3. Antioxidant Activity

Only one approach [122] is found for the influence of the toasting level on the antioxidant activity of the aged wine spirit. Surprisingly, in this work, it was observed that a non-significant variation of the DPPH inhibition with the toasting intensity existed (60.6%, LT; 69.6%, MT; 63.5%, ST) (Figure 8). Such an effect was assigned to the high variability associated with the toasting operation [102,164], despite the significant influence found for the majority of low molecular weight phenolic compounds (Table 5). Nevertheless, Híc et al. [168] noticed a similar behavior for the oak wood antioxidant activity under the toasting effect.

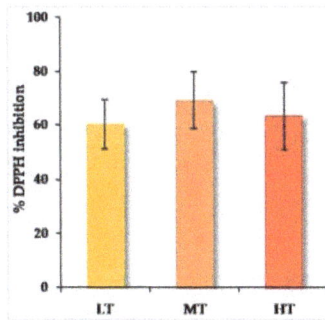

Figure 8. Mean values of antioxidant activity of the wine spirits aged in barrels with different toasting levels (LT—light toasting; MT—medium toasting; HT—heavy toasting). Adapted from [122].

5. Concluding Remarks

The reviewed literature demonstrates that the phenolic composition and some related properties of the aged wine spirit are effectively modulated by the kind of wood and the toasting level of the barrel. It is worth mentioning the highest enrichment of the wine spirit in low molecular weight phenolic compounds through the contact with chestnut wood (*Castanea sativa*). As a consequence, greater evolution of the chromatic characteristics and sensory color, as well as higher intensities of vanilla aroma, and higher antioxidant activity are achieved. *Q. pyrenaica* and *Quercus robur* exerts a similar influence. It means that these botanical species contribute to accelerate the ageing process and to give singular physicochemical characteristics and sensory profile to the aged wine spirit. The other kinds of oak (*Q. petraea* and *Q. alba*) show a weak performance, providing lower contents of phenolic compounds and promoting less intense related properties.

Regarding the toasting levels commonly used, higher concentrations of low molecular weight phenolic compounds, more evolved chromatic characteristics, sensory color, and other related sensory attributes are induced by the heavy toasting, followed by the medium toasting.

Thus, the wood botanical species together with the toasting level of the barrel are important resources for the industry for more sustainable management of the ageing process, to differentiate and to improve the quality of aged spirits. In addition, knowledge on the antioxidant activity of this beverage resulting from different ageing conditions may support a proper management of the ageing technology in order to add value to the final product. To be successful, the chemistry underlying the ageing process must be better understood. For this purpose, further research, supported by more advanced analytical methodologies, is needed on for key aspects such as: (1) identification and quantification of other phenolic compounds, coumarins and tannins of the aged wine spirit; (2) chemical reactions in which they are involved, and the relationship with chromatic characteristics and sensory properties of the aged wine spirit; (3) bioactive properties of the aged wine spirit modulated by thebarrel characteristics and other ageing factors. More studies about the heat treatment effect on the wood constituents and derived phenolic compounds, namely the coumarins, will also be of great relevance for a comprehensive insight into the ageing process.

Acknowledgments: The author thanks Professor Raúl Bruno de Sousa for his meaningful comments on this manuscript.

Conflicts of Interest: The author declares no conflicts of interest.

References

1. IBIS World Industry Report 2010. Global Spirits Manufacturing: C1122-GL. Available online: https://www.just-drinks.com/store/samples/2010_ibisworld (accessed on 27 August 2017).
2. Grigg, D. Wine, Spirits and Beer: World Patterns of Consumption. *Geography* **2004**, *89*, 99–110.
3. Léauté, R. Distillation in alambic. *Am. J. Enol. Vitic.* **1990**, *41*, 90–103.
4. Cantagrel, R. La qualité et le renom du Cognac dans le monde, sa place dans l'histoire. In *Les Eaux-De-Vie Traditionnelles D'origine Viticole*; Bertrand, A., Ed.; Lavoisier—Tec & Doc: Paris, France, 2008; pp. 15–38; ISBN 978-2-7430-1040-9.
5. Garreau, C. L'Armagnac. In *Les Eaux-De-Vie Traditionnelles D'origine Viticole*; Bertrand, A., Ed.; Lavoisier—Tec & Doc: Paris, France, 2008; pp. 39–62; ISBN 978-2-7430-1040-9.
6. Belchior, A.P.; Caldeira, I.; Costa, S.; Tralhão, G.; Ferrão, A.; Mateus, A.M.; Carvalho, E. Evolução das características fisico-químicas e organolépticas de aguardentes Lourinhã ao longo de cinco anos de envelhecimento em madeiras de carvalho e de castanheiro. *Ciência e Técnica Vitivinícola* **2001**, *16*, 81–94.
7. Regulation (EC) No. 110/2008. Definition, description, presentation, labelling and protection of geographical indications of spirit drinks. *Off. J. Eur. Union* **2008**, *L39*, 16–54.
8. Décret No. 2009-1285. Appellations d'origine contrôlée "Armagnac", "Blanche Armagnac", "Bas Armagnac", "Haut Armagnac" et "Armagnac-Ténarèze". *Journal Officiel de la République Française* **2009**, *247*, 17916–17927.
9. Décret No. 2009-1146. Appellation d'origine contrôlée "Cognac" ou "Eau-de-vie de Cognac" ou "Eau-de-vie des Charentes". *Journal Officiel de la République Française* **2009**, *221*, 15619–15628.
10. Decreto-Lei No. 323/94. Estatuto da Região Demarcada das Aguardentes Vínicas da Lourinhã. *Diário da República I Série A* **1994**, *29*, 7486–7489.
11. Caldeira, I.; Santos, R.; Ricardo-da-Silva, J.; Anjos, O.; Belchior, A.P.; Canas, S. Kinetics of odorant compounds in wine brandies aged in different systems. *Food Chem.* **2016**, *211*, 937–946. [CrossRef] [PubMed]
12. Canas, S. Aguardentes vínicas envelhecidas. In *Química Enológica—Métodos Analíticos. Avanços Recentes No Controlo da Qualidade de Vinhos e de Outros Produtos Vitivinícolas*; Curvelo-Garcia, A.S., Barros, P., Eds.; Publindústria, Edições Técnicas: Porto, Portugal, 2015; Capítulo 18.2; pp. 741–771; ISBN 978-989-723-118-6.

13. Baldwin, S.; Black, R.A.; Andreasen, A.A.; Adams, S.L. Aromatic congener formation in maturation of alcoholic distillates. *J. Agric. Food Chem.* **1967**, *15*, 381–385. [CrossRef]
14. Nishimura, K.; Ohnishi, M.; Masahiro, M.; Kunimasa, K.; Ryuichi, M. Reactions of wood components during maturation. In *Flavour of Distilled Beverages: Origin and Development*; Piggott, J.R., Ed.; Ellis Horwood Limited: Chichester, UK, 1983; pp. 241–255; ISBN 0-85312-546-5.
15. Puech, J.-L.; Leauté, R.; Clot, G.; Momdedeu, L.; Mondies, H. Évolution de divers constituants volatils et phénoliques des eaux-de-vie de Cognac au cours de leur vieillissement. *Sci. Aliments* **1984**, *4*, 65–80.
16. Nykanen, L. Formation and occurence of flavor compounds in wine and distilled alcoholic beverages. *Am. J. Enol. Vitic.* **1986**, *37*, 84–96.
17. Piggott, J.R.; Conner, J.M.; Clayne, J.; Peterson, A. The influence of non-volatile constituents on the extraction of ethyl esters from brandies. *J. Sci. Food Agric.* **1992**, *59*, 477–482. [CrossRef]
18. Singleton, V.L. Maturation of wines and spirits: Comparisons, facts and hypotheses. *Am. J. Enol. Vitic.* **1995**, *46*, 98–115.
19. Parke, S.A.; Birch, G.G. Solution properties of ethanol in water. *Food Chem.* **1999**, *67*, 241–246. [CrossRef]
20. Canas, S.; Belchior, A.P.; Mateus, A.M.; Spranger, M.I.; Bruno de Sousa, R. Kinetics of impregnation/evaporation and release of phenolic compounds from wood to brandy in experimental model. *Ciência e Técnica Vitivinícola* **2002**, *17*, 1–14.
21. Aronson, J.; Ebeler, S.E. Effect of polyphenol compounds on the headspace volatility of flavours. *Am. J. Enol. Vitic.* **2004**, *55*, 13–21.
22. Puech, J.-L.; Jouret, C.; Goffinet, B. Évolution des composés phénoliques du bois de chêne au cours du vieillissement de l'Armagnac. *Sci. Aliments* **1985**, *5*, 379–392.
23. Canas, S.; Belchior, A.P.; Caldeira, I.; Spranger, M.I.; Bruno de Sousa, R. La couleur et son évolution dans les eaux-de-vie *Lourinhã* pendant les trois premières années du vieillissement. *Ciência e Técnica Vitivinícola* **2000**, *15*, 1–14.
24. Caldeira, I.; Bruno de Sousa, R.; Belchior, A.P.; Clímaco, M.C. A sensory and chemical approach to the aroma of wooden aged Lourinhã wine brandy. *Ciência e Técnica Vitivinícola* **2008**, *23*, 97–110.
25. Belchior, A.P.; San-Romão, V. Influence de l'oxygène et de la lumière sur l'évolution de la composition phénolique des eaux-de-vie vieillis en bois de chêne. *Bull. Liaison Groupe Polyphenols* **1982**, *11*, 598–604.
26. Avakiants, S. Régulation des processus de vieillissement des eaux-de-vie. In *Élaboration et Connaissance des Spiritueux*; Cantagrel, R., Ed.; Lavoisier—Tec & Doc: Paris, France, 1992; pp. 595–600; ISBN 2-87777-3574.
27. Mosedale, J.R.; Puech, J.-L. Wood maturation of distilled beverages. *Trends Food Sci. Technol.* **1998**, *9*, 95–101. [CrossRef]
28. Canas, S.; Caldeira, I.; Belchior, A.P. Comparison of alternative systems for the ageing of wine brandy. Oxygenation and wood shape effect. *Ciência e Técnica Vitivinícola* **2009**, *24*, 33–40.
29. Cernîsev, S. Analysis of lignin-derived phenolic compounds and their transformations in aged wine distillates. *Food Control* **2017**, *73*, 281–290. [CrossRef]
30. Moutounet, M.; Mazauric, J.P.; Saint-Pierre, B.; Hanocq, J.F. Gaseous exchange in wines stored in barrels. *J. Sci. Tech. Tonnellerie* **1998**, *4*, 115–129.
31. Del Álamo-Sanza, M.; Nevares, I. Recent advances in the evaluation of the oxygen transfer rate in oak barrels. *J. Agric. Food Chem.* **2014**, *62*, 8892–8899. [CrossRef] [PubMed]
32. Del Álamo-Sanza, M.; Nevares, I.; Mayr, T.; Baro, J.A.; Martínez-Martínez, V.; Ehgartner, J. Analysis of the role of wood anatomy on oxygen diffusivity in barrel staves using luminescent imaging. *Sens. Actuators B Chem.* **2016**, *237*, 1035–1043. [CrossRef]
33. Canas, S.; Vaz, M.; Belchior, A.P. Influence de la dimension du fût dans les cinétiques d'extraction/oxydation des composés phénoliques du bois pour les eaux-de-vie Lourinhã. In *Les Eaux-de-vie Traditionnelles D'origine Viticole*; Bertrand, A., Ed.; Lavoisier—Tec & Doc: Paris, France, 2008; pp. 143–146. ISBN 978-2-7430-1040-9.
34. Philp, J.M. Cask quality and warehouse conditions. In *The Science and Technology of Whiskies*; Piggott, J.R., Sharp, R., Duncan, R.E.B., Eds.; Longman Scientific & Technical: Essex, UK, 1989; pp. 264–294; ISBN 978-0582044289.
35. Cantagrel, R.; Mazerolles, G.; Vidal, J.P.; Galy, B.; Boulesteix, J.M.; Lablanquie, O.; Gaschet, J. Evolution analytique et organoleptique des eaux-de-vie de Cognac au cours du vieillissement. 2ª partie: Incidence de la température et de l'hygrométrie des lieux de stockage. In *Élaboration et Connaissance des Spiritueux*; Cantagrel, R., Ed.; Lavoisier—Tec & Doc: Paris, France, 1992; pp. 573–576; ISBN 2-87777-3574.

36. Feuillat, F.; Perrin, J.R.; Keller, R. Simulation expérimentale del'interface tonneau. mesure des cinétiques d'imprégnation du liquide dans le bois et d'évaporation de surface. *OENO One* **1994**, *28*, 227–245. [CrossRef]

37. Puech, J.-L.; Léauté, R.; Mosedale, J.R.; Mourgues, J. Barrique et vieillissement des eaux-de-vie. In *Enologie Fondements Scientifiques et Technologiques*; Flanzy, C., Ed.; Lavoisier Tec & Doc: Paris, France, 1998; pp. 1110–1142; ISBN 978-2743002435.

38. Patrício, I.; Canas, S.; Belchior, A.P. Effect of brandies' agitation on the kinetics of extraction/oxidation and diffusion of wood extractable compounds in experimental model. *Ciência e Técnica Vitivinícola* **2005**, *20*, 1–15.

39. Belchior, A.P.; Carvalho, E. A cor em aguardentes vínicas envelhecidas: Método espectrofotométrico de determinação e relação com os teores em fenólicas totais. *Ciência e Técnica Vitivinícola* **1983**, *2*, 29–37.

40. Escolar, D.; Haro, M.; Saucedo, A.; Gòmes, J.; Alvàrez, J.A. Evolution de quelques paramètres physico-chimiques des brandies pendant leur vieillissement. *Doc. Blanc OIV* **1993**, *2023*, 1–9.

41. Bakker, J.; Bridle, P.; Timberlake, C.F. Tristimulus measurements (CIELab 76) of port wine colour. *Vitis* **1993**, *25*, 67–78.

42. Negueruela, A.I.; Echávarri, J.F.; Pérez, M.M. A study of correlation between enological colorimetric indexes and CIE colorimetric parameters in red wines. *Am. J. Enol. Vitic.* **1995**, *46*, 353–356.

43. Chassaing, S.; Lefeuvre, D.; Jacquet, R.; Jourdes, M.; Ducasse, L.; Galland, S.; Grelard, A.; Saucier, C.; Teissedre, P.-L.; Dangles, O.; et al. Physicochemical studies of new anthocyan-ellagitannin hybrid pigments: About the origin of the influence of oak C-glycosidic ellagitannins on wine color. *Eur. J. Org. Chem.* **2015**, *1*, 55–63.

44. Gambuti, A.; Capuano, R.; Lisanti, M.T.; Strollo, D.; Mioi, L. Effect of aging in new oak, one-year-used oak, chestnut barrels and bottle on color, phenolics and gustative profile of three monovarietal red wines. *Eur. Food Res. Technol.* **2010**, *231*, 455–465. [CrossRef]

45. Chinnici, F.; Natali, N.; Sonni, F.; Bellachiona, A.; Riponi, C. Comparative changes in color features and pigment composition of red wines aged in oak and cherry wood casks. *J. Agric. Food Chem.* **2011**, *59*, 6575–6582. [CrossRef] [PubMed]

46. Baiano, A.; De Gianni, A.; Mentana, A.; Quinto, M.; Centonze, D.; Del Nobile, M.A. Effects of the heat treatment with oak chips on color-related phenolics, volatile composition, and sensory profile of red wines: The case of Aglianico and Montepulciano. *Eur. Food Res. Technol.* **2016**, *242*, 745–767. [CrossRef]

47. Louw, L.; Oelofse, S.; Naes, T.; Lambrechts, M.; van Rensburg, P.; Nieuwoudt, H. Optimisation of the partial napping approach for the sucessful capturing of mouthfeel differentiation between brandy products. *Food Qual. Preference* **2015**, *41*, 245–253. [CrossRef]

48. Michel, J.; Albertin, W.; Jourdes, M.; Le Floch, A.; Giodanengo, T.; Mourey, N.; Teissedre, P.-L. Variations in oxygen and ellagitannins, and organoleptic properties of red wine aged in French oak barrels classified by a near infrared system. *Food Chem.* **2016**, *204*, 381–390. [CrossRef] [PubMed]

49. Canas, S.; Leandro, M.C.; Spranger, M.I.; Belchior, A.P. Phenolic compounds in a Lourinhã brandy extracted from different woods. In *Polyphenols Communications 98, Proceedings of the XIXth International Conference on Polyphenols, Lille, France, 1–4 September 1998*; Charbonnier, F., Delacotte, J.-M., Rolando, C., Eds.; Groupe Polyphénols: Bordeaux, France, 1998; Volume 1, pp. 373–374.

50. Lee, K.-Y.M.; Paterson, A.; Piggott, J.R.; Richardson, G.D. Origins of flavour in whiskies and revised flavour wheel: Review. *J. Inst. Brew.* **2001**, *107*, 287–313. [CrossRef]

51. Herve du Penhoat, C.L.M.; Michon, V.M.F.; Peng, S.; Viriot, C.; Scalbert, A.; Gage, D. Structural elucidation of new dimeric ellagitannins from *Quercus robur* L. Roburins A-E. *J. Chem. Soc. Perkin Trans.* **1993**, *I*, 1653–1660. [CrossRef]

52. Edelmann, A.; Lendl, B. Toward the optical tongue: Flow-through sensing of tannin-protein interactions based on FTIR spectroscopy. *J. Am. Chem. Soc.* **2002**, *124*, 14741–14747. [CrossRef] [PubMed]

53. Glabasnia, A.; Hofmann, T. Sensory-directed identification of taste-active ellagitannins in American (*Quercus alba* L.) and European oak wood (*Quercus robur* L.) and quantitative analysis in bourbon whiskey and oak-matured red wines. *J. Agric. Food Chem.* **2006**, *54*, 3380–3390. [CrossRef] [PubMed]

54. Hufnagel, J.C.; Hofmann, T. Orosensory-directed identification of astringent mouthfeel and bitter-tasting compounds in red wine. *J. Agric. Food Chem.* **2008**, *56*, 1376–1386. [CrossRef] [PubMed]

55. Marchal, A.; Cretin, B.N.; Sindt, L.; Waffo-Téguo, P.; Dubourdieu, D. Contribution of oak lignans to wine taste: Chemical identification, sensory characterization and quantification. *Tetrahedron* **2015**, *71*, 3148–3156. [CrossRef]

56. Sindt, L.; Gammacurta, M.; Waffo-Teguo, P.; Dubourdieu, D.; Marchal, A. Taste-guided isolation of bitter lignans from *Quercus petraea* and their identification in wine. *J. Nat. Prod.* **2016**, *79*, 2432–2438. [CrossRef] [PubMed]

57. Puech, J.-L. Vieillissement Des Eaux-de-vie en Futs de Chene. Extraction et Evolution de la Lignine et de ses Produits de Degradation. Ph.D. Thesis, Université Paul Sabatier de Toulouse, Toulouse, France, 1978.

58. Canas, S. Study of the Extractable Compounds of Woods (Oak and Chestnut) and the Extraction Processes in the Enological Perspective. Ph.D.Thesis, Instituto Superior de Agronomia, Universidade Técnica de Lisboa, Lisbon, Portugal, 2003.

59. Caldeira, I. The Aroma of Wine Brandies Aged in Wood. Cooperage Technology Relevance. Ph.D. Thesis, Instituto Superior de Agronomia, Universidade Técnica de Lisboa, Lisbon, Portugal, 2004.

60. Chen, J.H.; Ho, C.-T. Antioxidant activities of caffeic acid and its related hydroxycinnamic acid compounds. *J. Agric. Food Chem.* **1997**, *45*, 2374–2378. [CrossRef]

61. Priyadarsini, K.I.; Khopde, S.M.; Kumar, S.S.; Mohan, H. Free radical studies of ellagic acid, a natural phenolic antioxidant. *J. Agric. Food Chem.* **2002**, *50*, 2200–2206. [CrossRef] [PubMed]

62. Sroka, Z.; Cisowski, W. Hydrogen peroxide scavenging, antioxidant and anti-radical activity of some phenolic acids. *Food Chem. Toxicol.* **2003**, *41*, 753–758. [CrossRef]

63. Ou, S.; Kwok, K.C. Ferulic acid: Pharmaceutical functions, preparation and applications in foods. *J. Sci. Food Agric.* **2004**, *84*, 1261–1269. [CrossRef]

64. Soobrattee, M.A.; Neergheen, V.S.; Luximon-Ramma, A.; Arouma, O.I.; Bahorun, T. Phenolics as potential antioxidant therapeutic agents: Mechanisms and actions. *Mutat. Res.* **2005**, *579*, 200–213. [CrossRef] [PubMed]

65. Bakkalbasi, E.; Mentes, O.; Artik, N. Food—Occurence, effects of processing and storage. *Crit. Rev. Food Sci. Nutr.* **2009**, *49*, 283–298. [CrossRef] [PubMed]

66. Alañón, M.E.; Castro-Vázquez, L.C.; Diáz-Maroto, M.C.; Gordon, M.H.; Pérez-Coello, M.S. A study of the antioxidant capacity of oak wood used in wine ageing and the correlation with polyphenol composition. *Food Chem.* **2011**, *128*, 997–1002. [CrossRef]

67. Alañón, M.E.; Castro-Vázquez, L.; Díaz-Maroto, M.C.; Hermosín-Gutiérrez, I.; Gordon, M.H. Antioxidant capacity and phenolic composition of different woods used in cooperage. *Food Chem.* **2011**, *129*, 1584–1590. [CrossRef]

68. Kumar, K.N.; Raja, S.B.; Vidhya, N.; Devaraj, S.N. Ellagic acid modulates antioxidant status, ornithine decarboxylase expression, and aberrant crypt foci progression in 1,2-dimethylhydrazine-instigated colon preneoplastic lesions in rats. *J. Agric. Food Chem.* **2012**, *60*, 3665–3672. [CrossRef] [PubMed]

69. Heleno, S.A.; Martins, A.; Queiroz, M.J.R.P.; Ferreira, I.C.F.R. Bioactivity of phenolic acids: Metabolites versus parent compounds: A review. *Food Chem.* **2015**, *173*, 501–513. [CrossRef] [PubMed]

70. Bountagkidou, O.G.; Ordoudi, S.A.; Tsimidou, M.Z. Structure-antioxidant activity relationship of natural hydroxybenzaldehydes using in vitro assays. *Food Res. Int.* **2010**, *43*, 2014–2019. [CrossRef]

71. Ibrahim, M.N.M.; Sriprasanthi, R.B.; Shamsudeen, S.; Adam, F.; Bhawani, S.A. A concise review of the natural existance, synthesis, properties, and applications of syringaldehyde. *BioResources* **2012**, *7*, 4377–4399.

72. Skalicka-Wozniaka, K.; Erdogan Orhanb, I.A.; Cordellc, G.; Mohammad Nabavie, S.; Budzynska, B. Implication of coumarins towards central nervous system disorders. *Pharmacol. Res.* **2016**, *103*, 188–203. [CrossRef] [PubMed]

73. Hagerman, A.E.; Riedl, K.M.; Jones, G.A.; Sovik, K.N.; Ritchard, N.T.; Hartzfeld, P.W.; Riechel, T.L. High molecular weight plant polyphenolics (tannins) as biological antioxidants. *J. Agric. Food Chem.* **1998**, *46*, 1887–1892. [CrossRef] [PubMed]

74. Da Porto, C.; Calligaris, S.; Celotti, E.; Nicoli, M.C. Antiradical properties of commercial cognacs assessed by the DPPH test. *J. Agric. Food Chem.* **2000**, *48*, 4241–4245. [CrossRef] [PubMed]

75. Jordão, A.M.; Correia, A.C.; DelCampo, R.; SanJosé, M.L.G. Antioxidant capacity, scavenger activity, and ellagitannins content from commercial oak pieces used in winemaking. *Eur. Food Res. Technol.* **2012**, *235*, 817–825. [CrossRef]

76. Saleem, M.; Kim, J.H.; Alic, M.S.; Lee, Y.S. An update on bioactive plant lignans. *Nat. Prod. Rep.* **2005**, *22*, 696–716. [CrossRef] [PubMed]

77. Marteau, C.; Nardello-Rajat, V.; Favier, V.; Aubry, J.-M. Dual role of phenols as fragrances and antioxidants: Mechanism, kinetics and drastic solvent effect. *Flavour Frag. J.* **2013**, *28*, 30–38. [CrossRef]

78. Goldberg, D.M.; Hoffman, B.; Yang, J.; Soleas, G.J. Phenolic constituents, furans, and total antioxidant status of distilled spirits. *J. Agric. Food Chem.* **1999**, *47*, 3978–3985. [CrossRef] [PubMed]

79. Duriez, P.; Cren, C.; Luc, G.; Fruchart, J.C.; Rolando, C.; Teissier, E. Ingestion of cognac significantly increases plasma phenolic and ellagic acid concentrations and plasma antioxidant capacity in humans. In Proceedings of the 26th World Congress (81st Assembly of OIV)—Section Wine and Health, Adelaide, Australia, 11–17 October 2001; OIV: Adelaide, Australia, 2001; pp. 358–369.

80. Umar, A.; Boisseau, M.; Segur, M.-C.; Begaud, B.; Moore, N. Effect of age of Armagnac extract and duration of treatment on antithrombotic effects in a rat thrombosis model. *Thromb. Res.* **2003**, *111*, 185–189. [CrossRef] [PubMed]

81. Al Awwadi, N.A.; Borrot-Bouttefroy, A.; Umar, A.; Saucier, C.; Segur, M.-C.; Garreau, C.; Canal, M.; Glories, Y.; Moore, N. Effect of Armagnac fractions on human platelet aggregation in vitro and on rat arteriovenous shunt thrombosis in vivo probably not related only to polyphenols. *Thromb. Res.* **2007**, *119*, 407–412. [CrossRef] [PubMed]

82. Schwarz, M.; Rodríguez, M.C.; Martínez, C.; Bosquet, V.; Guillén, D.; Barroso, C.G. Antioxidant activity of Brandy de Jerez and other distillates, and correlation with their polyphenolic content. *Food Chem.* **2009**, *116*, 29–33. [CrossRef]

83. Ziyatdinova, G.; Salikhova, I.; Budnikov, H. Chronoamperometric estimation of cognac and brandy antioxidant capacity using MWNT modified glassy carbon electrode. *Talanta* **2014**, *125*, 378–384. [CrossRef] [PubMed]

84. Muresan, B.; Cimpoiu, C.; Hosu, A.; Bischin, C.; Gal, E.; Damian, G.; Fischer-Fodor, E.; Silaghi-Dumitrescu, R. Antioxidant content in Romanian traditional distilled alcoholic beverages. *Studia Ubb Chemia* **2015**, *60*, 355–370.

85. Garrido, J.; Borges, F. Wine and grape polyphenols—A chemical perspective. *Food Res. Int.* **2013**, *54*, 1844–1858. [CrossRef]

86. Botelho, G.; Canas, S.; Lameiras, J. Development of phenolic compounds encapsulation techniques as a major challenge for food industry and for health and nutrition fields. In *Nanotechnology in Agri-Food Industry*; Grumezescu, A.M., Ed.; Elsevier: London, UK, 2017; Chapter 14; Volume 5, pp. 535–586; ISBN 978-0-12-804304-2.

87. Deibner, L.; Jouret, C.; Puech, J.-L. Substances phénoliques des eaux-de-vie d'Armagnac. I. La lignin d'extraction et les produits de sa dégradation. *Ind. Aliment. Agricoles* **1976**, *93*, 401–414.

88. Puech, J.-L.; Jouret, C.; Deibner, L.; Alibert, G. Substances phénoliques des eaux-de-vie d'Armagnac et de rhum. II. Produits de la degradation de la lignin: Les aldéhydes et les acides aromatiques. *Ind. Aliment. Agricoles* **1977**, *94*, 483–493.

89. Puech, J.-L. Extraction and evolution of lignin products in Armagnac matured in oak. *Am. J. Enol. Vitic.* **1981**, *32*, 111–114.

90. Puech, J.-L.; Jouret, C. Dosage des aldéhydes aromatiques des eaux-de-vie conservées en fûts de chêne: Détection d'adultération. *Ann. Falsif. Exp. Chim.* **1982**, *805*, 81–90.

91. Belchior, A.P.; Puech, J.-L.; Carvalho, E.; Mondies, H. Caractéristiques de la composition phénolique du bois de chêne portugais et de quelques eaux-de-vie de vin. *Ciência e Técnica Vitivinícola* **1984**, *2*, 57–65.

92. Nabeta, K.; Yonekubo, J.; Miyake, M. Phenolic compounds from the heartwood of European oak (*Quercus robur* L.) and brandy. *Mokuzai Gakkaishi* **1987**, *33*, 408–415.

93. Salagoity-Auguste, M.H.; Tricard, C.; Sudraud, P. Dosage simultané des aldéhydes aromatiques et des coumarines par chromatographie liquide haute performance. *J. Chromatogr.* **1987**, *392*, 379–387. [CrossRef]

94. Tricard, C.; Salagoity, M.H.; Sudraud, P. La scopolétine: Un marqueur de la conservation en fûts de chêne. *OENO One* **1987**, *21*, 33–41. [CrossRef]

95. Puech, J.-L.; Moutounet, M. Liquid chromatographic determination of scopoletin in hydroalcoholic extract of oak wood and in matured distilled alcoholic beverages. *J. Assoc. Off. Anal. Chem.* **1988**, *71*, 512–514. [PubMed]

96. Artajona, J.; Barbero, E.; Llobet, M.; Marco, J.; Parente, F. Influence du "bousinage" de la barrique sur les qualités organoleptiques des brandies vieillies en fûts de chêne. In *Les Eaux-de-Vie Traditionnelles D'origine Viticole*; Bertrand, A., Ed.; Lavoisier—Tec & Doc: Paris, France, 1991; pp. 197–205; ISBN 2-85206-765-X.

97. Rabier, P.; Moutounet, M. Evolution d'extractibles de bois de chêne dans une eau-de-vie de vin. Incidence du thermotraitement des barriques. In *Les Eaux-de-vie Traditionnelles D'origine Viticole*; Bertrand, A., Ed.; Lavoisier—Tec & Doc: Paris, France, 1991; pp. 220–230; ISBN 2-85206-765-X.

98. Calvo, A.; Caumeil, M.; Pineau, J. Extraction des polyphénols et des aldéhydes aromatiques pendant le vieillissement du cognac, en fonction du titre alcoolique et du degré d'épuisement des fûts. In *Élaboration et Connaissance des Spiritueux*; Cantagrel, R., Ed.; Lavoisier—Tec & Doc: Paris, France, 1992; pp. 562–566; ISBN 2-87777-3574.

99. Puech, J.-L.; Moutounet, M. Phenolic compounds in an ethanol-water extract of oak wood and in brandy. *Food Sci. Technol.* **1992**, *25*, 350–352.

100. Puech, J.-L.; Maga, J. Influence du brûlage du fût sur la composition des substances volatiles et non volatiles d'une eau-de-vie. *Rev. Oenol.* **1993**, *70*, 13–16.

101. Viriot, C.; Scalbert, A.; Lapierre, C.; Moutounet, M. Ellagitannins and lignins in aging of spirits in oak barrels. *J. Agric. Food Chem.* **1993**, *41*, 1872–1879. [CrossRef]

102. Canas, S.; Leandro, M.C.; Spranger, M.I.; Belchior, A.P. Low molecular weight organic compounds of chestnut wood (*Castanea sativa* L.) and corresponding aged brandies. *J. Agric. Food Chem.* **1999**, *47*, 5023–5030. [CrossRef] [PubMed]

103. Panosyan, A.G.; Mamikonyan, G.V.; Torosyan, M.; Gabrielyan, E.S.; Mkhitaryan, S.A.; Tirakyan, M.R.; Ovanesyan, A. Determination of the composition of volatiles in Cognac (brandy) by headspace gas chromatography—Mass spectrometry. *J. Anal. Chem.* **2001**, *56*, 1078–1085. [CrossRef]

104. Savchuk, S.A.; Vlasov, V.N.; Appolonova, S.S.; Arbuzov, V.N.; Vedenin, A.N.; Mezinov, A.B.; Grigor'yan, B.R. Application of chromatography and spectrometry to the authentication of alcoholic beverages. *J. Anal. Chem.* **2001**, *56*, 214–231. [CrossRef]

105. Canas, S.; Belchior, A.P.; Spranger, M.I. Bruno de Sousa R. High-performance liquid chromatography method for analysis of phenolic acids, phenolic aldehydes and furanic derivatives in brandies. Development and validation. *J. Sep. Sci.* **2003**, *26*, 496–502. [CrossRef]

106. Canas, S.; Silva, V.; Belchior, A.P. Wood related chemical markers of aged wine brandies. *Ciência e Técnica Vitivinícola* **2008**, *23*, 45–52.

107. Cretin, B.N.; Dubourdieu, D.; Marchal, A. Development of a quantitation method to assay both lyoniresinol enatiomers in wines, spirits, and oak wood by liquid chromatography-high resolution mass spectrometry. *Anal. Bioanal. Chem.* **2016**, *408*, 3789–3799. [CrossRef] [PubMed]

108. Canas, S.; Leandro, M.C.; Spranger, M.I.; Belchior, A.P. Influence of botanical species and geographical origin on the content of low molecular weight phenolic compounds of woods used in Portuguese cooperage. *Holzforschung* **2000**, *54*, 255–261. [CrossRef]

109. Canas, S.; Grazina, N.; Spranger, M.I.; Belchior, A.P. Modelisation of heat treatment of Portuguese oak wood (*Quercus pyrenaica* Willd.). Analysis of the behaviour of low molecular weight phenolic compounds. *Ciência e Técnica Vitivinícola* **2000**, *15*, 75–94.

110. Canas, S.; Belchior, A.P.; Falcão, A.; Gonçalves, J.A.; Spranger, M.I.; Bruno de Sousa, R. Effect of heat treatment on the thermal and chemical modifications of oak and chestnut wood used in brandy ageing. *Ciência e Técnica Vitivinícola* **2007**, *22*, 5–14.

111. Alañón, M.E.; Rubio, H.; Díaz-Maroto, M.C.; Pérez-Coello, M.S. Monosaccharide anhydrides, new markers of toasted oak wood used for ageing wines and distillates. *Food Chem.* **2010**, *119*, 505–512. [CrossRef]

112. Sanz, M.; Cadahía, E.; Esteruelas, E.; Muñoz, A.M.; Fernández de Simón, B.; Hernández, T.; Estrella, I. Phenolic compounds in chestnut (*Castanea sativa* Mill.) heartwood. Effect of toasting at cooperage. *J. Agric. Food Chem.* **2010**, *58*, 9631–9640. [CrossRef] [PubMed]

113. Le Floch, A.; Jourdes, M.; Teissedre, P.-L. Polysaccharides and lignin from oak wood used in cooperage: Composition, interest, assays: A review. *Carbohydr. Res.* **2015**, *417*, 94–102. [CrossRef] [PubMed]

114. Hale, M.D.; McCafferty, K.; Larmie, E.; Newton, J.; Swan, J.S. The influence of oak seasoning and toasting parameters on the composition and quality of wine. *Am. J. Enol. Vitic.* **1999**, *50*, 495–502.

115. Acuña, L.; Gonzalez, D.; de la Fuente, J.; Moya, L. Influence of toasting treatment on permeability of six wood species for enological use. *Holzforschung* **2014**, *68*, 447–454. [CrossRef]

116. Puech, J.-L. Characteristics of oak wood and biochemical aspects of Armagnac aging. *Am. J. Enol. Vitic.* **1984**, *35*, 77–81.

117. Sarni, F.; Moutounet, M.; Puech, J.-L.; Rabier, P. Effect of heat treatment of oak wood extractable compounds. *Holzforschung* **1990**, *44*, 461–466. [CrossRef]
118. Chatonnet, P. Influence des Procédés de Tonnellerie et des Conditions D'élevage sur la Composition et la Qualité des Vins Elevés en Fûts de Chêne. Ph.D. Thesis, Institut d'Oenologie, Université de Bordeaux II, Bordeaux, France, 1995.
119. Cretin, B.N.; Sallembien, Q.; Sindt, L.; Daugey, N.; Buffeteau, T.; Waffo-Teguo, P.; Dubourdieu, D.; Marchal, A. How stereochemistry influences the taste of wine: Isolation, characterization and sensory evaluation of lyoniresinol stereoisomers. *Anal. Chim. Acta* **2015**, *888*, 191–198. [CrossRef] [PubMed]
120. Caldeira, I.; Clímaco, M.C.; Bruno de Sousa, R.; Belchior, A.P. Volatile composition of oak and chestnut woods used in brandy ageing: Modification induced by heat treatment. *J. Food Eng.* **2006**, *76*, 202–211. [CrossRef]
121. Chatonnet, P.; Boidron, J.N.; Pons, M. Incidence du traitement thermique du bois de chêne sur sa composition chimique. 2e partie: Évolution de certains composés en fonction de l'intensité de brûlage. Définition des paramètres thermiques de la chauffe des fûts en tonnellerie. *OENO One* **1989**, *4*, 223–250. [CrossRef]
122. Canas, S.; Casanova, V.; Belchior, A.P. Antioxidant activity and phenolic content of Portuguese wine aged brandies. *J. Food Compos. Anal.* **2008**, *21*, 626–633. [CrossRef]
123. Canas, S.; Spranger, M.I.; Belchior, A.P.; Bruno-de-Sousa, R. Isolation and identification by LC-ESI-MS of hydrolyzable tannins from *Quercus pyrenaica* Willd and *Castanea sativa* Mill heartwoods. In Proceedings of the 228th ACS National Meeting, Abstracts of the 4th Tannin Conference, Philadelphia, PA, USA, 22–26 August 2004; Gatenholm, P., Ed.; American Chemical Society: Philadelphia, PA, USA, 2004.
124. Fernández de Simón, B.F.; Sanz, M.; Cadahía, E.; Poveda, P.; Broto, M. Chemical characterization of oak heartwood from Spain forests of *Quercus pyrenaica* (Willd.) Ellagitannins, low molecular weight phenolic, and volatile compounds. *J. Agric. Food Chem.* **2006**, *54*, 8314–8321. [CrossRef] [PubMed]
125. Prida, A.; Boulet, J.-C.; Ducousso, A.; Nepveu, G.; Puech, J.-L. Effect of species and ecological conditions on ellagitannins content in oak wood from na even-aged and mixed stand of *Quercus robur* L. and *Quercus petraea* Liebl. *Ann. For. Sci.* **2006**, *63*, 415–424. [CrossRef]
126. Jordão, A.M.; Ricardo da Silva, J.M.; Laureano, O. Ellagitannins from Portuguese oak wood (*Quercus pyrenaica* Willd.) used in cooperage: Influence of geographical origin, coarseness of the grain and toasting level. *Holzforschung* **2007**, *61*, 155–160. [CrossRef]
127. Mammela, P.; Savolainen, H.; Lindroos, L.; Kangas, J.; Vartiainen, T. Analysis of oak tannins by liquid chromatography-electrospray ionisation mass spectrometry. *J. Chromatogr. A* **2001**, *891*, 75–83. [CrossRef]
128. Garcia, R.; Soares, B.; Dias, C.B.; Freitas, A.M.C.; Cabrita, M.J. Phenolic and furanic compounds of Portuguese chestnut and French, American and Portuguese oak wood chips. *Eur. Food Res. Technol.* **2012**, *235*, 457–467. [CrossRef]
129. Matricardi, L.; Waterhouse, A. Influence of toasting technique on color and ellagitannins of oak wood in barrel making. *Am. J. Enol. Vitic.* **1999**, *50*, 519–526.
130. Doussot, F.; De Jeso, B.; Quideau, S.; Pardon, P. Extractives content in cooperage oak wood during natural seasoning and toasting; influence of tree species, geographic location, and single-tree effects. *J. Agric. Food Chem.* **2002**, *50*, 5955–5961. [CrossRef] [PubMed]
131. Chira, K.; Teissedre, P.-L. Relation between volatile composition, ellagitannin content and sensory perception of oak wood chips representing different toasting processes. *Eur. Food Res. Technol.* **2013**, *236*, 735–746. [CrossRef]
132. García-Estévez, I.; Alcalde-Eon, C.; Le Grottaglie, L.; Rivas-Gonzalo, J.C.; Escribano-Bailón, M.T. Understanding the ellagitannin extraction process from oak wood. *Tetrahedron* **2015**, *71*, 3089–3094. [CrossRef]
133. Jourdes, M.; Michel, J.; Saucier, C.; Quideau, S.; Teissedre, P.-L. Identification, amounts, and kinetics of extraction of C-glucosidic ellagitannins during wine aging in oak barrels or in stainless steel tanks with oak chips. *Anal. Bioanal. Chem.* **2011**, *401*, 1531–1539. [CrossRef] [PubMed]
134. Lurton, L.; Ferrari, G.; Snakkers, G. Cognac: Production and aromatic characteristics. In *Alcoholic beverages. Sensory Evaluation and Consumer Research*; Piggott, J., Ed.; Woodhead Publishing Limited: Cambridge, UK, 2012; pp. 242–266. ISBN 978-0-85709-051-5.
135. Snakkers, G.; Nepveu, G.; Guilley, E.; Cantagrel, R. Variabilités géographique, sylvicole et individuelle de la teneur en extractibles de chênes sessiles français (*Quercus petraea* Liebl.): Polyphénols, octalactones et phénols volatils. *Ann. For. Sci.* **2000**, *57*, 251–260. [CrossRef]

136. Marchal, A.; Prida, A.; Dubourdieu, D. New approach for differentiating sessile and pedunculate oak: Development of a LC-HRMS method to quantitate triterpenoids in wood. *J. Agric. Food Chem.* **2016**, *64*, 618–628. [CrossRef] [PubMed]

137. Cadahía, E.; Munoz, L.; Simón, B.F.; Garcia-Vallejo, M.C. Changes in low molecular weight phenolic compounds in Spanish, French, and American oak woods during natural seasoning and toasting. *J. Agric. Food Chem.* **2001**, *49*, 1790–1798. [CrossRef] [PubMed]

138. Canas, S.; Caldeira, I.; Belchior, A.P.; Spranger, M.I.; Clímaco, M.C.; Bruno de Sousa, R. Chestnut wood: A sustainable alternative for the aging of wine brandies. In *Food Quality: Control, Analysis and Consumer Concerns*; Medina, D.A., Laine, A.M., Eds.; Nova Science Publishers Inc.: New York, NY, USA, 2011; pp. 181–228. ISBN 978-1-61122-917-2.

139. Canas, S.; Caldeira, I.; Mateus, A.M.; Belchior, A.P.; Clímaco, M.C.; Bruno de Sousa, R. Effect of natural seasoning on the chemical composition of chestnut wood used for barrel making. *Ciência e Técnica Vitivinícola* **2006**, *21*, 1–16.

140. De Rosso, M.; Cancian, D.; Panighel, A.; Vedova, A.D.; Flamini, R. Chemical compounds release from five different woods used to make barrels for aging wines and spirits. Volatile compounds and polyphénols. *Wood Sci. Technol.* **2009**, *43*, 375–385. [CrossRef]

141. Castellari, M.; Piermattei, B.; Arfelli, G.; Amati, A. Influence of aging conditions on the quality of red sangiovese wine. *J. Agric. Food Chem.* **2001**, *49*, 3672–3676. [CrossRef] [PubMed]

142. Alañón, M.E.; Schumacher, R.; Castro-Vásquez, L.; Díaz-Maroto, M.C.; Hermosín-Gutiérrez, I.; Pérez-Coello, M.S. Enological potential of chestnut wood for aging Tempranillo wines. Part II: Phenolic compounds and chromatic characteristics. *Food Res. Int.* **2013**, *51*, 536–543. [CrossRef]

143. Rodríguez Madrera, R.; Suárez Valles, B.; Diñeiro García, Y.; del Valle Arguelles, P.; Picinelli Lobo, A. Alternative woods for aging distillates—An insight into their phenolic profiles and antioxidant activities. *Food Sci. Biotechnol.* **2010**, *19*, 1129–1134. [CrossRef]

144. Canas, S.; Quaresma, H.; Belchior, A.P.; Spranger, M.I.; Bruno de Sousa, R. Evaluation of wine brandies authenticity by the relationships between benzoic and cinnamic aldehydes and between furanic aldehydes. *Ciência e Técnica Vitivinícola* **2004**, *19*, 13–27.

145. Doussot, F.; Pardon, P.; Dedier, J.; De Jeso, B. Individual, species and geographic origin influence on cooperage oak extractible content (*Quercus robur* L. and *Quercus petraea* Liebl.). *Analusis* **2000**, *28*, 960–965. [CrossRef]

146. Quaglieri, C.; Jourdes, M.; Waffo-Teguo, P.; Teissedre, P.-L. Updated knowledge about pyranoanthocyanins: Impact of oxygen on their contents, and contribution in the winemaking process to overall wine color. *Trends Food Sci. Technol.* **2017**, *67*, 139–149. [CrossRef]

147. Christensen, C.M. Effects of colour on aroma, flavour and texture judgements of foods. *J. Food Sci.* **1983**, *48*, 787–790. [CrossRef]

148. Schwarz, M.; Rodríguez, M.C.; Guillén, D.A.; Barroso, C.G. Analytical charactisation of a Brandy de Jerez during its ageing. *Eur. Food Res. Technol.* **2011**, *232*, 813–819. [CrossRef]

149. Rodríguez-Solana, R.; Salgado, J.M.; Domínguez, J.M.; Cortés-Diéguez, S. First approach to the analytical characterization of barrel-aged grape marc distillates using phenolic compounds and colour parameters. *Food Technol. Biotechnol.* **2014**, *52*, 391–402. [CrossRef] [PubMed]

150. Es-Safi, N.; Cheynier, V.; Moutounet, M. Study of the reactions between (+)-catechin and furfural derivatives in the presence or absence of anthocyanins and their implication in food color change. *J. Agric. Food Chem.* **2002**, *48*, 5946–5954. [CrossRef]

151. Es-Safi, N.; Cheynier, V.; Moutounet, M. Role of aldehydes in the condensation of phenolic compounds with emphasis on food organoleptic properties. *J. Agric. Food Chem.* **2002**, *50*, 5571–5585. [CrossRef] [PubMed]

152. Sousa, C.; Mateus, N.; Perez-Alonso, J.; Santos-Buelga, C.; De Freitas, V. Preliminary study of oaklins, a new class of brick-red catechinpyrylium pigments resulting from the reaction between catechin and wood aldehydes. *J. Agric. Food Chem.* **2005**, *53*, 9249–9256. [CrossRef] [PubMed]

153. Caldeira, I.; Mateus, A.M.; Belchior, A.P. Flavour and odour profile modifications during the first five years of Lourinhã brandy maturation on different wooden barrels. *Anal. Chim. Acta* **2006**, *563*, 264–273. [CrossRef]

154. Renaud, S.; De Lorgeril, M. Wine, alcohol, platelets and The French Paradox for coronary heart disease. *Lancet* **1992**, *339*, 1523–1526. [CrossRef]

155. Tunstall-Pedoe, H.; Kuulasmaa, K.; Mahonen, M.; Tolonen, H.; Ruokokoski, E.; Amouyed, P. Contribution of trends in survival and coronary-event rates to changes in coronary heart disease mortality: 10-year results from 37 WHO MONICA project populations. Monitoring trends and determinants in cardiovascular disease. *Lancet* **1999**, *353*, 1547–1557. [CrossRef]

156. Alonso, A.M.; Castro, R.; Rodriguez, M.C.; Guillen, D.A.; Barroso, C.G. Study of the antioxidant power of brandies and vinegars derived from Sherry wines and correlation with their content in polyphenols. *Food Res. Int.* **2004**, *37*, 715–721. [CrossRef]

157. Pecic, S.; Veljovic, M.; Despotovic, S.; Leskosek-Cukalovic, I.; Jadranin, M.; Tesevic, V.; Niksic, M.; Nikicevic, N. Effect of maturation conditions on sensory and antioxidant properties of old Serbian plum brandies. *Eur. Food Res.Technol.* **2012**, *235*, 479–487. [CrossRef]

158. Psarra, E.; Makris, D.P.; Kallithraka, S.; Kefalas, P. Evaluation of the antiradical and reducing properties of selected Greek white wines: Correlation with polyphenolic composition. *J. Sci. Food Agric.* **2002**, *82*, 1014–1020. [CrossRef]

159. Ávila-Reyes, J.; Almarz-Abarca, N.; Delgado-Alvarado, E.A.; González-Valdez, L.; Del Toro, G.V.; Páramo, E.D. Phenol profile and antioxidant capacity of mescal aged in oak wood barrels. *Food Res. Int.* **2010**, *43*, 296–300. [CrossRef]

160. Yeh, C.-T.; Yen, G.-C. Modulation of hepatic phase II phenol sulfotransferase and antioxidant status by phenolic acids in rats. *J. Nutr. Biochem.* **2006**, *17*, 561–569. [CrossRef] [PubMed]

161. Chatonnet, P.; Boindron, J.N. Incidence du traitement thermique du bois de chêne sur sa composition chimique. 1ere partie: Définition des paramètres thermiques de la chauffe des fûts en tonnellerie. *OENO One* **1989**, *23*, 77–87. [CrossRef]

162. Sanz, M.; Fernández de Simón, B.; Cadahía, E.; Esteruelas, E.; Muñoz, A.M.; Hernández, M.T.; Estrella, I. Polyphenolic profile as a useful tool to identify the wood used in wine aging. *Anal. Chim. Acta* **2012**, *732*, 33–45. [CrossRef] [PubMed]

163. Duval, C.J.; Sok, N.; Laroche, J.; Gourrat, K.; Prida, A.; Lequin, S.; Chassagne, D.; Gougeon, R.D. Dry vs soaked wood: Modulating the volatile extractible fraction of oak wood by heat treatments. *Food Chem.* **2013**, *138*, 270–277. [CrossRef] [PubMed]

164. Cantagrel, R.; Mazerolles, G.; Vidal, J.P.; Galy, B. Evolution analytique et organoleptique des eaux-de-vie de cognac au cours du vieillissement. 1ère partie: Incidence des techniques de tonnelleries. In *Élaboration et Connaissance des Spiritueux*; Cantagrel, R., Ed.; Lavoisier—Tec & Doc: Paris, France, 1992; pp. 567–572; ISBN 2-87777-3574.

165. Puech, J.-L.; Lepoutre, J.-P.; Baumes, R.; Bayonove, C.; Moutounet, M. Influence du thermotraitement des barriques sur l'évolution de quelques composants issus du bois de chêne dans les eaux-de-vie. In *Élaboration et Connaissance des Spiritueux*; Cantagrel, R., Ed.; Lavoisier—Tec & Doc: Paris, France, 1992; pp. 583–594; ISBN 2-87777-3574.

166. Sanz, M.; Cadahia, E.; Esteruelas, E.; Muñoz, A.M.; Fernández de Simon, B.; Hernández, T.; Estrella, I. Effect of toasting intensity at cooperage on phenolic compounds in acacia (*Robinia pseudoacacia*) heartwood. *J. Agric. Food Chem.* **2011**, *59*, 3135–3145. [CrossRef] [PubMed]

167. Chira, K.; Teissedre, P.-L. Chemical and sensory evaluation of wine matured in oak barrel: Effect of oak species involved and toasting process. *Eur. Food Res. Technol.* **2015**, *240*, 533–547. [CrossRef]

168. Híc, P.; Soural, I.; Balík, J.; kulichová, J.; Vrchotová, N.; Tríska, J. Antioxidant capacities of extracts in relation to toasting oak and acacia wood. *J. Food Nutr. Res.* **2017**, *56*, 129–137.

beverages

MDPI

Review

Recent Advances and Applications of Pulsed Electric Fields (PEF) to Improve Polyphenol Extraction and Color Release during Red Winemaking

Arianna Ricci, Giuseppina P. Parpinello * [iD] and Andrea Versari [iD]

Department of Agricultural and Food Sciences, Alma Mater Studiorum—University of Bologna,
Piazza Goidanich 60, 47521 Cesena, FC, Italy; arianna.ricci4@unibo.it (A.R.); andrea.versari@unibo.it (A.V.)

Academic Editors: António Manuel Jordão and James Harbertson
Received: 30 November 2017; Accepted: 13 January 2018; Published: 1 March 2018

Abstract: Pulsed electric fields (PEF) technology is an innovative food processing system and it has been introduced in relatively recent times as a pre-treatment of liquid and semi-solid food. Low cost-equipment and short processing time, coupled to the effectiveness in assisting the extraction of valuable compounds from vegetable tissues, makes PEF a challenging solution for the industrial red winemaking; a tailored PEF-assisted maceration was demonstrated to promote an increase in wine color quality and an improvement in the polyphenolic profile. Despite the application of PEF has been studied and the positive effects in selected wine varieties were demonstrated on batch and pilot-scale systems, there is a need for a more detailed characterization of the impact in different grapes, and for a better understanding of potential undesirable side-effects. This review aims to summarize the state of the art in view of a detailed feasibility study, to promote the introduction of PEF technology in the oenological industry.

Keywords: pulsed electric field; polyphenolic extraction; red winemaking; color intensity

1. Introduction

The implementation of food products in terms of nutritional value and shelf-durability and the optimization of the production processes (with a consequent reduction of the production costs) have been the main challenges faced by the food industry in recent decades. Based on these premises, innovative technologies with potential application on an industrial scale have been investigated with the aim of enhancing food quality, increasing competitiveness in the food market and matching costumer's expectation [1–3].

Thermal processing is traditionally used for the biological stabilization, by subjecting the food to a temperature range from 60 to above 100 °C for variable periods, and involving a massive energy transfer from the heat source to the treated matrix [4]. Although the energy obtained is effective in destroying or inhibiting undesirable microorganisms, many unwanted secondary reactions leading to the loss of nutritional and sensorial quality of food have been highlighted in recent studies [5,6]. Mild thermal treatments are also applied to enhance mass transfer phenomena, obtaining the removal of water from food and the release of nutraceutical compounds; despite the temperature supply is limited compared to the pasteurization treatment, the use of heat in fresh food matrices potentially results in the deterioration of sensory properties [7–9].

Non-thermal food processing involving ambient temperatures constitute a concrete alternative to thermal technologies, improving food safety while maintaining product quality and economic feasibility; several non-thermal technologies were proposed for preserving the nutritional constituents of food including vitamins, minerals, and essential flavors. High hydrostatic pressures, oscillating magnetic fields, intense light pulses, irradiation, the use of chemicals and biochemicals, high-intensity pulsed electric fields were all recognized as emerging solutions in the early 2000s [10,11]. Among them, pulsed electric field (PEF) represents a promising emerging solution for the food industry, being effective on the process scale and competitive in terms of cost. The use of pulsed electric field was introduced in the food industry in the early 90*ies*, and it is based on the application of high-frequencies electric pulses with intense field strengths, which are able to modulate the activity of biological membranes. The use of high-voltage pulsed treatments was initially hypothesized for inactivation of microorganisms and pasteurization of liquid foods [12]; afterwards, the improved understanding on the mechanisms involved, and the development of efficient pilot-scale equipment with reduced processing time and continuous flow applicability have extended the application fields in the food industry [13–15].

The "dielectric breakdown" theory provides a reliable mechanistic description of the impact of electric pulses in the modification of biological membranes. Accordingly, the membrane of a cell is modeled as a capacitor filled with a low-dielectric constant fluid; after being exposed to a strong electric field, ions migrate along the fluid and toward the membrane walls, thus forming free charges of opposite sign which accumulate at both membrane sides. Due to the attractive interaction of charges, the cell wall undergoes a compression which reduces the membrane thickness; the mechanism proceeds until the electric field strength gains a critical value, usually around 1V potential, at which micropores are formed on the membrane, increasing permeability (electroporation). The electroporated membrane may be damaged owing to the direct rupture, as a consequence of Joule overheating of the membrane surface, or following the chemical imbalances caused by the enhanced transmembrane transport throughout the membrane pores.

In a typical PEF treatment microsecond, intense electric pulses are applied to a conductive material (whether it is liquid or solid) which is located between two working electrodes; the external electric field induces a reversible or irreversible permeabilization of biological cells membranes in organic food matrices, supporting step production processes like tissues soaking, peels removal or the extraction of bioactive compounds [12,16–18].

The mechanism of alteration of transmembrane permeability has also gained specific interest for the oenological industry, having a potential application for improving the maceration stage in red vinification. As the majority of the compounds responsible for the quality and stability of wine color (polyphenolic compounds, colour precursors, tannins) are located in the grape skin vacuoles of the grape berries, the mass transfer is a critical mechanism for achieving their efficient extraction, and the winemaking techniques aim to increase the permeability of cytoplasmic membranes to facilitate their release; however, an effective extraction of such compounds strongly impacts the industrial process in terms of duration and cost. Different techniques have been developed to enhance the extraction of polyphenols and pigmented compounds improving the performances of the static maceration: thermovinification, grape freezing, the use of maceration enzyme, among others, have been tested on pilot and industrial scales [19–21]. To date, many of these approaches require a significant amount of energy and cause losses of valuable nutraceuticals: conducting alcoholic fermentation at high temperatures can cause fermentation failures and loss of volatile compounds, freezing grapes was demonstrated to affect the wine quality on different extents, and commercial formulations of maceration enzymes provide different levels of purity and potential contamination of detrimental species like β-glucosidases [22–24].

Treating red grapes with pulsed fields is a challenging opportunity to enhance the mass transfer phenomenon reducing the duration of the extraction process, thus limiting the economic impact and potential side effects of maceration. The bulk of research on the oenological field has evolved over the years 2007–2012, with few advances in the following years; during this period, along with the efficacy of the treatment, attention was paid to the potential undesirable effects, which have inhibited the application of PEF on an industrial scale so far [25].

The aim of this review is a critical evaluation of the state of art, to overcome actual limitations and promote the application of PEF technology in the wine industry. The manuscript is structured in the following sections: (i) generic overview on the PEF technology; (ii) a brief description of the

PEF-assisted mass transfer of valuable compounds from vegetable tissues; (iii) presentation of the main scientific contributions to the use of the PEF technology in winemaking, considering the decade 2007–2017, and also highlighting limitations and potential challenges.

2. An Overview on PEF Technology: Equipment and Processes

PEF involves the direct application of very short current voltage pulses to the food matrices; the time duration of the pulses cycling is variable (ranging micro to milliseconds) during which the material placed between two electrodes is treated with high voltages; the specific intensity of the treatment could be modulated on the basis of the geometry and distance of the working electrodes, the voltage delivered, the conductivity of the material treated. There is no strict definition regulating such intensity: on the basis of empirical experience, treatment ranges required to increase polyphenols and color extraction have been defined as high ($E > 1$ kV cm^{-1}) medium ($E \approx 0.1$–1 kV cm^{-1}) and low-intensity ($E < 0.1$ kV cm^{-1}) electric fields [26].

The basic components of a typical PEF apparatus for producing exponentially decay pulses are schematically represented in Figure 1: regardless the settings required for specific industrial processes, the bulk of the equipment is constituted by a generator of electric pulses and a treatment chamber. In the generator, a charger converts the AC to DC current supplying an energy storage device, and a switch shortly turn on and off the high voltage circuit to generate electric pulses; the discharge of electrical energy is a critical step due to high voltages and short timing involved, and the capacitor is continuously monitored and stepped up if voltage collapse [27]. For square pulses, the electric scheme is more complex involving several LRC circuits (electrical circuits consisting of a resistor (R), an inductor (L), and a capacitor (C)) associated. The treatment chamber is composed by two electrodes separated by isolating materials and a gap to be filled with the food to be treated; the distance between electrodes, the voltage applied and the geometry of the chamber affect the treatment by defining the strength of the electric field, and modulating the energy supplied per unit of area. The food industry treatments frequently take place under continuous flow, thus making difficult modelling and predicting the distribution of the electric field along the treated material; most preliminary studies have provided static models were the electrodes have coaxial, co-linear (non-uniform electric field distribution) or parallel flat (uniform electric field distribution) electrodes, and the scale-up provides laminar flow at constant rates to optimize the treatment's efficiency [28].

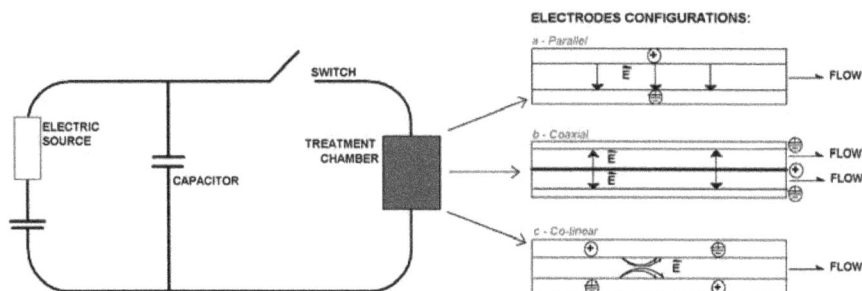

Figure 1. Schematic representation of a pulsed electric fields (PEF) circuit system for producing exponentially decay pulses, including the structure of the most exploited treatment chambers.

After modulating the set-up of the apparatus, the following process parameters must be defined: (i) electric field strength; (ii) pulse shape; (iii) pulse width; (iv) number of pulses; (v) pulse specific energy, (vi) pulse frequency [27]. The electric field strength is also related to the uniformity of electric field, which is dependent on the specific set-up of the apparatus as previously mentioned: in the parallel configuration of electrodes the treatment is uniform along the treated matrix, while in collinear and co-axial set-ups intensity gradients of the electric field are formed within the treatment chamber

(Figure 1). The commonly used pulse shapes are exponential decays and square waveform, the latter being the most suitable to maximize the effect of a PEF treatment: in square waveform any power decay occurs and the intensity remains constant for the whole duration pulse width, increasing efficiency compared to the pulses sent per unit of time. The specific energy delivered during the treatment is a function of the voltage applied, duration of the treatment, resistance and conductivity of the material treated, and temperature variation induced by electric field in the material, and it constitute a process parameter that can be modulated based on preliminary trials; the parameters described are also used to evaluate the PEF treatment efficiency and economic impact in prevision of scale-up [29].

3. Enhancement of Mass Transfer from Vegetable Tissues by PEF

Electroporation is a physical mechanism to induce permeabilization of cell membranes through the application of external intense electric pulses, and it can be exploited in the food processing [30]. In more detail, the application of a PEF treatment in plant cell could assist the extraction of valuable compounds such as pigments, antioxidants, flavors, contained in membranes and vacuoles of plant tissues. In eukaryote plant cells permeabilization of the membrane is easier to obtain if compared with bacterial cell, resulting in lower electric intensities required and subsequent lower energy consumption; for this reason, PEF treatment is considered competitive in term of cost/effectiveness for industrial treatments like pressing, extraction, drying and diffusion [31]. The reversible electroporation or the electrical membrane breakdown are due to the intrinsic composition and electric properties of cell membranes, which induce amplification of the external electric field. More specifically, the conductivity of the intact cell membrane is several orders of magnitude lower than that of the medium and cytoplasm where it is immersed. When an external electric field is applied, the opposite charges migrate and accumulate at the interface of membrane with the medium, increasing the transmembrane potential. Due to the thickness of the a typical vegetal cell membrane ($\approx$5 nm), which is very low compared to a typical plant cell radius ($\approx$100 μm), the electric field accumulates distributing opposite charges at the two interfaces (inner/outer) of the membrane itself; the electrostatic attraction of opposite charges operates along the membrane, inducing a squeezing of tissues. When the electric field reaches a critical value (Ec), ranging 0.2–1 V/m for vegetal cells, the compression induces a reversible (E $\approx$ Ec) or irreversible (E >> Ec) formation of pores [32,33].

In Figure 2 the time course of the electroporation process is schematically represented, showing that it is a non-instantaneous, dynamic process. The kinetic of electroporation is the reason for the key role played by technological parameters such as the intensity, duration and the pulse shape in a typical PEF treatment; moreover, the degree of degradation of cell tissues influences the kinetic of extraction of desired compounds.

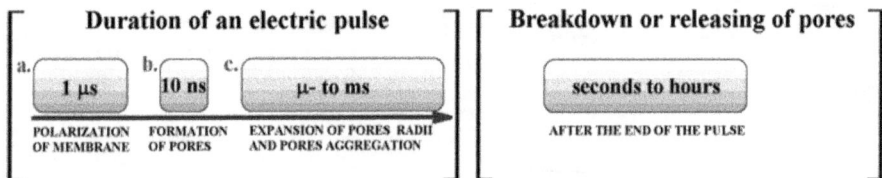

Figure 2. Time-course of a typical PEF experiment after releasing pulses on a generic eukaryote vegetable cells.

4. PEF Technology in Red Winemaking (Decade 2007–2017)

Grapes skin contain a large number of phenolic compounds and natural pigments, which are protected in the cell walls and cytoplasmic membranes of the skin vacuoles, and have high technological relevance for the oenological industry; nevertheless, due to the presence of cellular barriers that increase their resistance to the mass transfer, they are only partially (and slowly) extracted and released in wine during traditional maceration processes that usually lasts up to 2–3 weeks. The main phenolic compounds contained in red grapes are anthocyanins, polyphenolic monomers, condensed tannins and pigmented polymers, all contributing to the color quality and stability. The final content and composition of polyphenols in wine is affected by several factors like grape variety, cultivation factor, winemaking technique, efficient extraction yield of the maceration step. For instance, the extraction of wine phenolics can be enhanced using technological alternatives to traditional maceration, but each of them has showed limitations in terms of cost-effectiveness and potentially have a negative impact on the sensory profile of wines. PEF has been recently introduced as an alternative pre-maceration treatment to increase and speed-up polyphenolic extraction without altering the sensory properties, highlighting effectiveness in improving wine stability and color quality [34]. An early-stage trial conducted using a PEF-assisted juice expression of red grapes resulted in an increasing anthocyanins content juice compared to non-thermal processes, such as high pressure treatment, supercritical CO_2, or the subjection to ultrasounds [35]. In 2007, the experiment was extended to Muscadelle, Sauvignon and Semillon white grape varieties, with the aim of increasing the expression and quality of juice: the PEF-assisted expression provided an increase of juice yield from 50% to 80% following a 5 bar pressing for 45 min [36]. Following the improvement in grape juice quality, an increasing number of experiment were conducted in the years 2008–2009, aimed to monitor not only the improvement of juice expression, but also the possibility to increase the polyphenols extraction in grape juice. The PEF-assisted low-pressure mechanical expression (maximum pressure: 1 bar) was also tested to assay Chardonnay white grapes, increasing juice yield from 67% to 75%, and producing a 15% increase of the polyphenolic content; moreover, the visual aspect of the juice was improved while reducing the juice turbidity with respect to the control [37].

The promising preliminary results on grape juices have stimulated an increasing number of experiments aimed to applying PEF prior to the maceration step of red wines. López et al. [38] have investigated the effects of pre-treating Tempranillo grape skin with pulsed electric field treatments (5 and 10 kV/cm) on the total polyphenols and anthocyanins content, color attributes. It was observed a generic increase and improvement of the three parameters when the electric energy irradiation raised from 5 to 10 kV/cm. After 50 h maceration the maximum index of polyphenols and anthocyanins content was obtained in PEF-treated samples, while the same value was obtained after 96 h in the control. Moreover, the total polyphenols index in PEF treated wines was 13.7% higher (5 kV/cm treatment) and 29.0% higher (10 kV/cm treatment) with respect to the control after 96 h maceration. At the end of fermentation the Tempranillo treated wines showed a color intensity of 23.93 (control), 27.04 (5 kV/cm treatment), and 29.33 (10 kV/cm treatment), confirming the color improvement in induced by PEF pre-treatment [38]. The same Authors have investigate the influence of the grape variety on the effectiveness of PEF treatments in red wine, performing the pulsed electric field on autochthonous Spanish grape varieties (Garnacha, Mazuelo and Graciano). Irradiation treatments of 2, 5 and 10 kV/cm were applied to improve color intensity, anthocyanins and total polyphenols index, guaranteeing a low energy consumption to obtain cell permeabilization of grape skins (0.4–6.7 kJ/kg) and short processing times. The PEF treatment was more effective in Mazuelo variety than in Garnacha and Graciano, reaching color intensity of 21.2% (2 kV/cm), 35.3% (5 kV/cm), 49.8% (10 kV/cm) and anthocyanins 20.3% (2 kV/cm), 28.6% (5 kV/cm), 41.8% (10 kV/cm) higher than the control after 120 h maceration [39]. Nevertheless, according to a study after Luengo et al. that was conducted few years later, in 2014, the PEF treatment can be modulated to achieve optimal polyphenols extraction and color expression in Garnacha red wine [40].

Tempranillo, Graciano and Garnacha Spanish grape varieties were also used as a template to demonstrate that a PEF treatment induce an increase in the extraction yield of *trans*-resversatrol and piceid-like compounds (stilbenes), which have a major antioxidant and nutraceutical impact but are usually available in low concentrations in red wines, being difficult to release from grape tissues using traditional extraction methods [41].

Cabernet sauvignon is a native grape variety from the Bordeaux region, which has been widespread in different geographical areas due to its ductility and excellent sensorial properties. It produces intensely colored wines, with high contents in tannins and flavors, suitable for aging; long maceration times and aging in oak barrels could enhance the sensory properties of pure or blended Cabernet sauvignon wines. Due to its peculiar properties, Cabernet sauvignon has been one of the main grape variety assayed to evaluate the advantages of PEF-assisted winemaking.

The intermediate 5 kV/cm PEF treatment was applied to the grape pomace of Cabernet Sauvignon, with an irradiation energy of 2.1 kJ/kg. The aim of the experiment was to evaluate the evolution on quality parameters and anthocyanins content on finished wine obtained after different maceration times: 48, 72, 96, 248 h, using PEF as a pre-treatment, and comparing with the same wine without applying electrical pulses to the grape juice. The PEF treatment of Cabernet sauvignon grapes was confirmed to produce intensely colored wines at bottling, with shorter maceration times: evaluations concerning the PEF experiment have highlighted the possibility of reducing the maceration time from 268 h (control) to 72 h (PEF treatment), obtaining comparable anthocyanins and polyphenols extraction yields [42]. Nevertheless, a major concern regarded the evolution of polyphenolic and color profile throughout aging in bottles. For this reason, the same grape variety was used to make a comparative study between PEF-assisted maceration of the grape mass pumped through a co-linear treatment chamber, and the use of 2 commercial enzymatic preparations in Cabernet sauvignon grapes under static conditions. PEF treatment conditions were defined on the basis of previous experiment after Lopez et al., and consisted of 50 pulses of 5 kV/cm at a frequency of 122 Hz, resulting in a total specific energy of 3.67 kJ/kg. The following parameters were monitored: color intensity (CI), anthocyanins content (AC) and total polyphenol index (TPI), from grape crushing to 3 months of wine aging in bottle. After 3 months, both enzymatic preparations were increasing the CI level of 5% with respect to the control, and only one of them reached the maximum increase of 11% (AC) and 3% (TPI). PEF was evaluated as the best alternative, with higher increase of CI, AC, and TPI parameters (28%, 26%, and 11%, respectively), compared to the control. The HPLC analysis also highlighted the improved extraction of non-anthocyanins polyphenols in PEF-assisted maceration [43].

A further experiment after Puértolas et al. [44] provided the application of a PEF treatment (50 pulses, 122 Hz frequency, 5 kV/cm electric field strength, 3.67 kJ/kg total specific energy) before crushing of the Cabernet sauvignon grapes; musts obtained from PEF-treated grapes were fermented concurrently with untreated grape must, and the evolution of the two bottled wines during 12 months of storage was compared in terms of color intensity, anthocyanins, polyphenols as measured using the Folin–Ciocalteu method. Assuming that the monitored parameters were higher on the PEF-treated sample, it was found that aging did not affect the initial color intensity in both wines, whereas the Folin–Ciocalteu index decreased. Individual polyphenolic compounds showed similar evolution patterns during the 12 months storage, and the content of flavonoids and hydroxycinnamic acids, which are valuable compounds for improving wine aging, was constantly higher in PEF wines [44].

The main challenge in the application PEF for improving winemaking provides the design of a device operating in continuous flow; such a configuration would make feasible the introduction of the PEF technology on the oenological industry scale. A continuous flow treatment should avoid the backfilling and compaction of grape pomaces to be treated, to provide a constant flow steam and ensure homogeneous distribution of electric pulses. Puértolas et al. [45] have developed a pilot-scale system operating in continuous flow, powered by a progressive cavity pump to obtain a grape mass flow rate of 118 kg/h which ensure a resident time in the treatment zone of 0.41 s/unit of mass. The device was equipped with a co-linear treatment chamber; the co-linear design provided two treatment zones of 2 cm between the electrodes, with the inner diameter of 2 cm. The simulation of non-uniform distribution of the electric field in co-linear configuration (see Figure 1) was supported by the Comsol Multiphysics software (Comsol Inc., Stockholm, Sweden). PEF parameters were consistent with the treatments previously applied in Cabernet sauvignon to obtain an enhancement in color and extractability [42,44], obtaining a moderate increase in the temperature of the treated mass

(+2 °C maximum). Results of the PEF pilot-scale experiment confirmed previous achievements of experiment conducted in batch systems of low capacity [45].

An ultimate study on Cabernet sauvignon berries was conducted comparing PEF treatment of low strength, long duration and high energy (0.7 kV/cm, 200 ms, 31 Wh/kg) with high-strength treatment (4 kV/cm, 1 ms, 4 Wh/kg); the experiment was mainly focused on the evaluation of amounts and composition of extractable anthocyanins and tannins. In particular, the first treatment have highlighted a major capacity in extracting the parietal tannins of the skin (+34%), while the second showed higher effectiveness in anthocyanins extraction (+19%) and the resulting must was richer in vacuolar tannins. In both cases, a depolymerisation of high-molecular weight fraction of skin tannins was observed, improving the diffusion of small polymers along the cell membrane pores. Enological parameters were monitored and they do not exhibit significant modifications; however, the sensory impact in terms of astringency produced by low molecular weight tannins has not been tested in this work [46].

The same authors previously investigated the effect of pulsed electric field treatments with variable energies (0.5–0.7 kV/cm) and duration (40–100 ms) of the treatment in Merlot grapes. Kinetical models provided the evolution of total polyphenols, anthocyanins and color intensities content in juices, while keeping the temperature variation below 5 °C. The increase in extractability was demonstrated using kinetic parameters, and the sensory evaluation of treated wines evidenced an improvement in the sensory properties of PEF-treated wines compared to the control [47].

The combination between pulsed electric field treatment and oenological practices to maximize color quality in wine has been investigated in a study after Puértolas et al. The experimental design was aimed to study the effect of maceration time (0–6 h), and maceration temperature (4–20 °C) on the anthocyanin content during rose vinification of Cabernet sauvignon grapes, while keeping constant parameters for the PEF treatment (5 kV/cm, 50 pulses). A surface response model analysis was used to identify optimal process parameters, showing that after 2 h treatment the PEF treatment allowed to obtain an anthocyanin level of 50 mg/L at the temperature of 4 °C; the same result was obtained for the control (untreated) sample using a 20 °C temperature. Since an increasing processing temperature results in the speed-up of oxidation mechanisms and loss of flavors, the PEF treatment showed to be a promising technique to minimize temperature in rose vinification, avoiding undesirable side-effects [48].

The use of PEF to improve polyphenols, anthocyanins and tannins extraction for Cabernet franc grapes was investigated by El Darra et al. comparing moderate (0.8 kV/cm, 42 kJ/kg) and high (5 kV/cm, 53 kJ/kg) intensities PEF treatments with moderate thermal (MT) and ultrasound-assisted (US) extractions. Despite all pre-treatments have provided improved polyphenolic and color profiles compared to the control, PEF treatments showed the high extraction yields of total polyphenols (+51% for moderate and +62% for high-intensity treatments, respectively), and the highest color intensities (+20% for moderate and +23% for high-intensity treatments, respectively) at the end of maceration [49].

According to our knowledge only two autochthonous Italian varieties, Aglianico and Piedirosso grapes, were subjected to PEF treatments to increase the polyphenolic content of related red wines, and different electrical field strength were applied (0.5–1–5 kV/cm) obtaining specific energy inputs ranging from 1 to 50 kJ/kg. It was observed that PEF had a major effect on Aglianico, increasing the release of polyphenols (+20%), anthocyanins (+75%), color intensity (+20%), and enhancing the antioxidant activity of the wine (+20%); contrariwise, same treatments had a minor impact in Piedirosso composition in comparison to the control [50].

The contribution of PEF treatment coupled to conventional red wine fining techniques has also been investigated. The evolution of chromatic and phenolic profiles of Cabernet Sauvignon during aging in oak barrels and following bottling was studied, showing that the stabilising effect of applying pulsed field prior must fermentation is maintained and even enhanced during aging in American oak barrels: in the study, a panel triangle test was used to demonstrate that the wine quality remained unchanged in terms of color and phenolic content after 8 months bottling, regardless the bottling was preceded by storage in wood under oxidative conditions [51].

Based on the results of published studies, we can conclude that the mass diffusion of valuable polyphenols, anthocyanins, tannins and pigments is improved using PEF-assisted maceration, in different extent when modulating engineering (field strength, specific energy, treatment time) and technological parameters (grape variety, winemaking technology, storage time), and these findings could support both red winemaking and the valorization of production wastes obtained during the red vinification [52]. Table 1 summarizes the main findings of studies listed in this review, to obtain a more generic overview of PEF applications which have showed to improve the red winemaking practice so far.

Table 1. Summary of the main findings in PEF experiments in relation to grape varieties and process parameters. Abbreviation: Total Polyphenols Index (TPI), Color Intensity (CI), Anthocyanins Content (AC).

Grape Variety	PEF Parameters	Major Achievements (Compared to Control Wine)	Ref.
Mazuelo	(a) 2 kV/cm; 0.4 kJ/kg	TPI + 19.8%; CI + 21.2%; AC + 20.3% (120 h maceration)	[39]
	(b) 5 kV/cm; 1.8 kJ/kg	TPI + 24.0%; CI + 35.3%; AC + 28.6% (120 h maceration)	
	(c) 10 kV/cm; 6.7 kJ/kg	TPI + 31.0%; CI + 49.8%; AC + 41.8% (120 h maceration)	
Tempranillo	(a) 5 kV/cm; 1.8 kJ/kg	TPI + 13.7%; CI + 11.5%; AC + 21.5% (96 h maceration)	[38]
	(b) 10 kV/cm; 6.7 kJ/kg	TPI + 29.0%; CI + 18.4%; AC + 28.6% (96 h maceration)	
Garnacha	4.3 kV/cm	TPI + 23.5%; CI + 12.5%; AC + 25% (7 days maceration)	[40]
Cabernet sauvignon	5 kV/cm; 2.1 kJ/kg	TPI + 45.2%; CI + 48.4%; AC + 42.2% (268 h maceration)	[42]
Cabernet sauvignon	5 kV/cm; 3.67 kJ/kg	TPI + 17%; CI + 28% (12 months aging); AC + 17% (end of maceration)	[43]
Cabernet sauvignon	5 kV/cm; 3.67 kJ/kg	TPI + 23%; CI + 29%; AC + 26% (4 months aging)	[44]
Cabernet sauvignon	5 kV/cm; 3.67 kJ/kg	TPI + 23%; CI + 34% (48 h maceration); CI + 38% (4 months aging)	[45]
Cabernet sauvignon	5 kV/cm; 3.67 kJ/kg	Decrease in maceration T from 20 °C to 4 °C to have comparable results with control	[48]
Cabernet sauvignon	(a) 0.7 kV/cm; 31 Wh/kg	Tannins extraction + 19%	[46]
	(b) 4 kV/cm; 4 Wh/kg	Anthocyanins extraction + 19%	
Merlot	(a) 0.7 kV/cm; 40 ms	TPI + 18% (all)	[47]
	(b) 0.7 kV/cm; 100 ms	CI—no significant differences	
	(c) 0.5 kV/cm; 100 ms	Increasing kinetic of extraction with increasing process parameters	
Cabernet franc	(a) 0.8 kV/cm; 42 kJ/kg	TPI + 51%; CI + 20%; AC + 49%	[49]
	(b) 5 kV/cm; 53 kJ/kg	TPI + 62%; CI + 23%; AC + 60%	
Aglianico	(a) 1 kV/cm; 50 kJ/kg	TPI + 13%; CI + 6%; AC + 9% (following treatment)	[50]
	(b) 1.5 kV/cm; 10 kJ/kg	TPI + 31%; CI + 12%; AC + 54% (following treatment)	
	(c) 1.5 kV/cm; 25 kJ/kg	TPI + 38%; CI + 19%; AC + 76% (following treatment)	

Regardless the obvious advantages in PEF-assisted winemaking, the development of pulsed electric fields in the oenological industry have been limited due to the limited understanding of the impact of this treatment on wine composition and quality parameters. A recent review after Yang et al. highlighted some concerning related to the use of PEF in the food industry: in particular, it has been proven that the release of metals under the effect of the external electric field is the main latent effect of PEF. Corrosion and migration of electrodes material can occur under specific conditions, and this could result in food contamination and adulteration [53]. As an example, a study conducted in beer has shown a significant increase in Fe, Cr, Zn, Mn concentrations, which are oxidation catalysts; in this study, the impact of increasing metals concentration in the flavor of beer was determined by the sensory panel [54]. Contrariwise, any sensory analysis conducted on PEF-treated wines has highlighted detrimental modifications in comparison with the control wine, but there is a lack in a detailed study of elemental and compositional modifications occurring in wines after the application of electric treatments. A further negative effect following the application of PEF in wine could be the selective electrochemical degradation of valuable chemical compounds, depending on the electrodes used. A recent study compared the effect of three electrodes materials in aqueous solution of anthocyanin-based compounds, showing that stainless steel retained cyanidin-3-glucoside and cyanidin-3-sophoroside, while pure titanium and titanium-based alloy caused a higher extent of degradation [55]. Lastly, the corrosion or migration of electrode materials should be investigated to improve safety and prevent undesirable sensorial modifications in wines.

A further observation regards the limited number of publications made available in recent times, the reduced number of technological parameters monitored, and the limited selection of grape varieties investigated. According to the information made available so far, the authors of this review consider pulsed electric field a challenging solution for improving red winemaking, to be deepened and extended to several case studies for a better understanding of electrochemical processes involved and a safe exploitation of the technique in industrial applications.

5. Conclusions

The most recent findings in the field of red wine processing using the pulsed electric field (PEF) technology have been summarized in this review. It has been demonstrated that the use of PEF could enhance the extraction of phenolic compounds and increase color intensity in red wines. In the studies presented, color stability was also assisted by the extraction of valuable polyphenolic compounds, monomers (flavonoids, hydroxycinnamic acids, tannins) and polymers (condensed tannins from grape skin). The main mechanism responsible for the increasing mass transfer was the electroporation of vegetable cell membranes, which also enabled to reduce the maceration process, and to enhance the extraction of bioactive compounds from grape using a low-cost technology. Regardless the advantages reported for the application of PEF in red winemaking, the impact of the treatment in the overall composition of wine has not been investigated in detail so far, and a major concern regards the potential electrochemical contaminations induced by the electrodes. Moreover, a limited selection of grape varieties was investigated, and there is a need for a more comprehensive study about the impact of PEF treatments in the chemical and sensory properties of different red wines. The authors encourage a more systematic study of PEF technique applied to winemaking, for improving safety and promoting the development of this technology on an industrial scale.

Acknowledgments: Author A.R. is supported by a post-doc fellowship from the University of Bologna, Italy.

Author Contributions: Author A.R. wrote the paper and provided the theoretical explanation of PEF technology; authors G.P.P. and A.V. provided the critical review of PEF experiments in winemaking.

Conflicts of Interest: The authors declare no conflict of interest.

References

1. Vilkhu, K.; Mawson, R.; Simons, L.; Bates, D. Applications and opportunities for ultrasound assisted extraction in the food industry—A review. *Innov. Food Sci. Emerg. Technol.* **2008**, *9*, 161–169. [CrossRef]
2. Costa, A.I.A.; Dekker, M.; Jongen, W.M.F. Quality function deployment in the food industry: A review. *Trends Food Sci. Technol.* **2000**, *11*, 306–314. [CrossRef]
3. Cardello, A.V.; Schutz, H.G.; Lesher, L.L. Consumer perceptions of foods processed by innovative and emerging technologies: A conjoint analytic study. *Innov. Food Sci. Emerg. Technol.* **2007**, *8*, 73–83. [CrossRef]
4. Jay, J.M. Microbiological food safety. *Crit. Rev. Food Sci. Nutr.* **1992**, *31*, 177–190. [CrossRef] [PubMed]
5. Alwazeer, D.; Delbeau, C.; Divies, C.; Cachon, R. Use of redox potential modification by gas improves microbial quality, color retention, and ascorbic acid stability of pasteurized orange juice. *Int. J. Food Microbiol.* **2003**, *89*, 21–29. [CrossRef]
6. Torregrosa, F.; Cortés, C.; Esteve, M.J.; Frígola, A. Effect of high-intensity pulsed electric fields processing and conventional heat treatment on orange–carrot juice carotenoids. *J. Agric. Food Chem.* **2005**, *53*, 9519–9525. [CrossRef] [PubMed]
7. Kim, Y.S.; Park, S.J.; Cho, Y.H.; Park, J. Effects of combined treatment of high hydrostatic pressure and mild heat on the quality of carrot juice. *J. Food Sci.* **2001**, *66*, 1355–1360. [CrossRef]
8. Loginova, K.V.; Vorobiev, E.; Bals, O.; Lebovka, N.I. Pilot study of countercurrent cold and mild heat extraction of sugar from sugar beets, assisted by pulsed electric fields. *J. Food Eng.* **2011**, *102*, 340–347. [CrossRef]
9. Lebovka, N.I.; Praporscic, I.; Vorobiev, E. Effect of moderate thermal and pulsed electric field treatments on textural properties of carrots, potatoes and apples. *Innov. Food Sci. Emerg. Technol.* **2004**, *5*, 9–16. [CrossRef]
10. Butz, P.; Tauscher, B. Emerging technologies: Chemical aspects. *Food Res. Int.* **2002**, *35*, 279–284. [CrossRef]
11. Morris, C.; Brody, A.L.; Wicker, L. Non-thermal food processing/preservation technologies: A review with packaging implications. *Packag. Technol. Sci.* **2007**, *20*, 275–286. [CrossRef]

12. Barbosa-Canovas, G.V.; Pierson, M.D.; Zhang, Q.H.; Schaffner, D.W. Pulsed Electric Fields. *J. Food Sci.* **2000**, *65*, 65–79. [CrossRef]

13. Góngora-Nieto, M.M.; Sepúlveda, D.R.; Pedrow, P.; Barbosa-Cánovas, G.V.; Swanson, B.G. Food processing by pulsed electric fields: Treatment delivery, inactivation level, and regulatory aspects. *Food Sci. Technol.* **2002**, *35*, 375–388. [CrossRef]

14. Zhao, W.; Yu, Z.; Liu, J.; Yu, Y.; Yin, Y.; Lin, S.; Chen, F. Optimized extraction of polysaccharides from corn silk by pulsed electric field and response surface quadratic design. *J. Sci. Food Agric.* **2011**, *91*, 2201–2209. [CrossRef] [PubMed]

15. Corrales, M.; Toepfl, S.; Butz, P.; Knorr, D.; Tauscher, B. Extraction of anthocyanins from grape by-products assisted by ultrasonics, high hydrostatic pressure or pulsed electric fields: A comparison. *Innov. Food Sci. Emerg. Technol.* **2008**, *9*, 85–91. [CrossRef]

16. Tylewicz, U.; Tappi, S.; Mannozzi, C.; Romani, S.; Dellarosa, N.; Laghi, L.; Ragni, L.; Rocculi, P.; Dalla Rosa, M. Effect of pulsed electric field (PEF) pre-treatment coupled with osmotic dehydration on physico-chemical characteristics of organic strawberries. *J. Food Eng.* **2017**, *213*, 2–9. [CrossRef]

17. Dellarosa, N.; Tappi, S.; Ragni, L.; Laghi, L.; Rocculi, P.; Dalla Rosa, M. Metabolic response of fresh-cut apples induced by pulsed electric fields. *Innov. Food Sci. Emerg. Technol.* **2016**, *38*, 356–364. [CrossRef]

18. Traffano-Schiffo, M.V.; Laghi, L.; Castro-Giraldez, M.; Tylewicz, U.; Rocculi, P.; Ragni, L.; Dalla Rosa, M.; Fito, P.J. Osmotic dehydration of organic kiwifruit pre-treated by pulsed electric fields and monitored by NMR. *Food Chem.* **2017**, *236*, 87–93. [CrossRef] [PubMed]

19. Sacchi, K.L.; Bisson, L.F.; Adams, D.O. A review of the effect of winemaking techniques on phenolic extraction in red wines. *Am. J. Enol. Vitic.* **2005**, *56*, 197–206.

20. Pinelo, M.; Arnous, A.; Meyer, A.S. Upgrading of grape skins: Significance of plant cell-wall structural components and extraction techniques for phenol release. *Trends Food Sci. Technol.* **2006**, *17*, 579–590. [CrossRef]

21. Aguilar, T.; Loyola, C.; de Bruijn, J.; Bustamante, L.; Vergara, C.; von Baer, D.; Mardones, C.; Serra, I. Effect of thermomaceration and enzymatic maceration on phenolic compounds of grape must enriched by grape pomace, vine leaves and canes. *Eur. Food Res. Technol.* **2016**, *242*, 1149–1158. [CrossRef]

22. Bautista-Ortín, A.B.; Martínez-Cutillas, A.; Ros-García, J.M.; López-Roca, J.M.; Gómez-Plaza, E. Improving colour extraction and stability in red wines: The use of maceration enzymes and enological tannins. *Int. J. Food Sci. Technol.* **2005**, *40*, 867–878. [CrossRef]

23. Hüfner, E.; Haßelbeck, G. Application of Microbial Enzymes During Winemaking. In *Biology of Microorganisms on Grapes, in Must and in Wine*; König, H., Unden, G., Fröhlich, J., Eds.; Springer: Cham, Germany, 2017; pp. 635–658.

24. Geffroy, O.; Siebert, T.; Silvano, A.; Herderich, M. Impact of winemaking techniques on classical enological parameters and rotundone in red wine at the laboratory scale. *Am. J. Enol. Vitic.* **2017**, *68*, 141–146. [CrossRef]

25. Puértolas, E.; López, N.; Condón, S.; Álvarez, I.; Raso, J. Potential applications of PEF to improve red wine quality. *Trends Food Sci. Technol.* **2010**, *21*, 247–255. [CrossRef]

26. Asavasanti, S.; Ersus, S.; Ristenpart, W.; Stroeve, P.; Barrett, D.M. Critical electric field strengths of onion tissues treated by pulsed electric fields. *J. Food Sci.* **2010**, *75*, E433–E443. [CrossRef] [PubMed]

27. Puértolas, E.; Luengo, E.; Álvarez, I.; Raso, J. Improving mass transfer to soften tissues by pulsed electric fields: Fundamentals and applications. *Annu. Rev. Food Sci. Technol.* **2012**, *3*, 263–282. [CrossRef] [PubMed]

28. Huang, K.; Wang, J. Designs of pulsed electric fields treatment chambers for liquid foods pasteurization process: A review. *J. Food Eng.* **2009**, *95*, 227–239. [CrossRef]

29. Heinz, V.; Alvarez, I.; Angersbach, A.; Knorr, D. Preservation of liquid foods by high intensity pulsed electric fields—basic concepts for process design. *Trends Food Sci. Technol.* **2001**, *12*, 103–111. [CrossRef]

30. Tsong, T.Y. Electroporation of cell membranes. *Biophys. J.* **1991**, *60*, 297–306. [CrossRef]

31. Knorr, D.; Angersbach, A.; Eshtiaghi, M.N.; Heinz, V.; Lee, D.-U. Processing concepts based on high intensity electric field pulses. *Trends Food Sci. Technol.* **2001**, *12*, 129–135. [CrossRef]

32. Vorobiev, E.; Lebovka, N. Pulsed-electric-fields-induced effects in plant tissues: Fundamental aspects and perspectives of applications. In *Electrotechnologies for Extraction from Food Plants and Biomaterials*; Springer: New York, NY, USA, 2008; pp. 39–81.

33. Donsì, F.; Ferrari, G.; Pataro, G. Applications of pulsed electric field treatments for the enhancement of mass transfer from vegetable tissue. *Food Eng. Rev.* **2010**, *2*, 109–130. [CrossRef]

34. Morata, A.; Loira, I.; Vejarano, R.; González, C.; Callejo, M.J.; Suárez-Lepe, J.A. Emerging preservation technologies in grapes for winemaking. *Trends Food Sci. Technol.* **2017**, *67*, 36–43. [CrossRef]

35. Knorr, D. Impact of non-thermal processing on plant metabolites. *J. Food Eng.* **2003**, *56*, 131–134. [CrossRef]

36. Praporscic, I.; Lebovka, N.; Vorobiev, E.; Mietton-Peuchot, M. Pulsed electric field enhanced expression and juice quality of white grapes. *Sep. Purif. Technol.* **2007**, *52*, 520–526. [CrossRef]

37. Grimi, N.; Lebovka, N.I.; Vorobiev, E.; Vaxelaire, J. Effect of a pulsed electric field treatment on expression behavior and juice quality of Chardonnay grape. *Food Biophys.* **2009**, *4*, 191–198. [CrossRef]

38. López, N.; Puértolas, E.; Condón, S.; Álvarez, I.; Raso, J. Effects of pulsed electric fields on the extraction of phenolic compounds during the fermentation of must of Tempranillo grapes. *Innov. Food Sci. Emerg. Technol.* **2008**, *9*, 477–482. [CrossRef]

39. López, N.; Puértolas, E.; Condón, S.; Álvarez, I.; Raso, J. Application of pulsed electric fields for improving the maceration process during vinification of red wine: Influence of grape variety. *Eur. Food Res. Technol.* **2008**, *227*, 1099–1107. [CrossRef]

40. Luengo, E.; Franco, E.; Ballesteros, F.; Álvarez, I.; Raso, J. Winery trial on application of pulsed electric fields for improving vinification of Garnacha grapes. *Food Bioprocess Technol.* **2014**, *7*, 1457–1464. [CrossRef]

41. López-Alfaro, I.; González-Arenzana, L.; López, N.; Santamaría, P.; López, R.; Garde-Cerdán, T. Pulsed electric field treatment enhanced stilbene content in Graciano, Tempranillo and Grenache grape varieties. *Food Chem.* **2013**, *141*, 3759–3765. [CrossRef] [PubMed]

42. López, N.; Puértolas, E.; Hernández-Orte, P.; Álvarez, I.; Raso, J. Effect of a pulsed electric field treatment on the anthocyanins composition and other quality parameters of Cabernet Sauvignon freshly fermented model wines obtained after different maceration times. *Food Sci. Technol.* **2009**, *42*, 1225–1231. [CrossRef]

43. Puértolas, E.; Saldaña, G.; Condón, S.; Álvarez, I.; Raso, J. A comparison of the effect of macerating enzymes and pulsed electric fields technology on phenolic content and color of red wine. *J. Food Sci.* **2009**, *74*, C647–C652. [CrossRef] [PubMed]

44. Puértolas, E.; Saldaña, G.; Condón, S.; Álvarez, I.; Raso, J. Evolution of polyphenolic compounds in red wine from Cabernet Sauvignon grapes processed by pulsed electric fields during aging in bottle. *Food Chem.* **2010**, *119*, 1063–1070. [CrossRef]

45. Puértolas, E.; Hernández-Orte, P.; Sladaña, G.; Álvarez, I.; Raso, J. Improvement of winemaking process using pulsed electric fields at pilot-plant scale. Evolution of chromatic parameters and phenolic content of Cabernet Sauvignon red wines. *Food Res. Int.* **2010**, *43*, 761–766. [CrossRef]

46. Delsart, C.; Cholet, C.; Ghidossi, R.; Grimi, N.; Gontier, E.; Gény, L.; Vorobiev, E.; Mietton-Peuchot, M. Effects of pulsed electric fields on Cabernet Sauvignon grape berries and on the characteristics of wines. *Food Bioprocess Technol.* **2014**, *7*, 424–436. [CrossRef]

47. Delsart, C.; Ghidossi, R.; Poupot, C.; Cholet, C.; Grimi, N.; Vorobiev, E.; Milisic, V.; Mietton Peuchot, M. Enhanced extraction of phenolic compounds from Merlot grapes by pulsed electric field treatment. *Am. J. Enol. Vitic.* **2012**, *63*, 205–211. [CrossRef]

48. Puértolas, E.; Saldaña, G.; Álvarez, I.; Raso, J. Experimental design approach for the evaluation of anthocyanin content of rosé wines obtained by pulsed electric fields. Influence of temperature and time of maceration. *Food Chem.* **2011**, *126*, 1482–1487. [CrossRef]

49. El Darra, N.; Grimi, N.; Maroun, R.G.; Louka, N.; Vorobiev, E. Pulsed electric field, ultrasound, and thermal pretreatments for better phenolic extraction during red fermentation. *Eur. Food Res. Technol.* **2013**, *236*, 47–56. [CrossRef]

50. Donsì, F.; Ferrari, G.; Fruilo, M.; Pataro, G. Pulsed electric field-assisted vinification of Aglianico and Piedirosso grapes. *J. Agric. Food Chem.* **2010**, *58*, 11606–11615. [CrossRef] [PubMed]

51. Puértolas, E.; Saldaña, G.; Alvarez, I.; Raso, J. Effect of pulsed electric field processing of red grapes on wine chromatic and phenolic characteristics during aging in oak barrels. *J. Agric. Food Chem.* **2010**, *58*, 2351–2357. [CrossRef] [PubMed]

52. Castro-López, C.; Rojas, R.; Sánchez-Alejo, E.J.; Niño-Medina, G.; Martínez-Ávila, G.C.G. Phenolic compounds recovery from grape fruit and by-products: An overview of extraction methods. In *Grape and Wine Biotechnology*; Morata, A., Ed.; InTech: London, UK, 2016; pp. 103–123.

53. Yang, N.; Huang, K.; Lyu, C.; Wang, J. Pulsed electric field technology in the manufacturing processes of wine, beer, and rice wine: A review. *Food Control* **2016**, *61*, 28–38. [CrossRef]

54. Evrendilek, G.A.; Li, S.; Dantzer, W.R.; Zhang, Q.H. Pulsed electric field processing of beer: Microbial, sensory, and quality analyses. *J. Food Sci.* **2004**, *69*, M228–M232. [CrossRef]
55. Sun, J.; Bai, W.; Zhang, Y.; Liao, X.; Hu, X. Effects of electrode materials on the degradation, spectral characteristics, visual colour, and antioxidant capacity of cyanidin-3-glucoside and cyanidin-3-sophoroside during pulsed electric field (PEF) treatment. *Food Chem.* **2011**, *128*, 742–747. [CrossRef]

beverages

MDPI

Review

Phenolic Compounds and Antioxidant Activity in Grape Juices: A Chemical and Sensory View

Fernanda Cosme [1,*], Teresa Pinto [2] and Alice Vilela [1] (ID)

[1] Chemistry Research Centre (CQ-VR), Department of Biology and Environment, School of Life Sciences and Environment, University of Trás-os-Montes and Alto Douro, Edifício de Enologia, Apartado 1013, 5001-801 Vila Real, Portugal; avimoura@utad.pt

[2] Centre for the Research and Technology of Agro-Environmental and Biological Sciences (CITAB), Department of Biology and Environment, School of Life Sciences and Environment, University of Trás-os-Montes and Alto Douro, 5000-801 Vila Real, Portugal; tpinto@utad.pt

Received: 13 December 2017; Accepted: 1 March 2018; Published: 6 March 2018

Abstract: The search for food products that promote health has grown over the years. Phenolic compounds present in grapes and in their derivatives, such as grape juices, represent today a broad area of research, given the benefits that they have on the human health. Grape juice can be produced from any grape variety once it has attained appropriate maturity. However, only in traditional wine producing regions, grape juices are produced from *Vitis vinifera* grape varieties. For example, Brazilian grape juices are essentially produced from *Vitis labrusca* grape varieties, known as American or hybrid, as they preserve their characteristics such as the natural flavour after pasteurisation. Grapes are one of the richest sources of phenolic compounds among fruits. Therefore, grape juices have been broadly studied due to their composition in phenolic compounds and their potential beneficial effects on human health, specifically the ability to prevent various diseases associated with oxidative stress, including cancers, cardiovascular and neurodegenerative diseases. Therefore, this review will address grape juices phenolic composition, with a special focus on the potential beneficial effects on human health and on the grape juice sensory impact.

Keywords: grape juice; phenolic compounds; antioxidant activity; bioactive compounds; sensory analysis

1. General Introduction

Grape juice is a clear or cloudy liquid extracted from the grapes using different technological processes, being the main techniques employed "Hot press" (HP), "Cold press" (CP) and "Hot Break" (HB). It is a non-fermented, non-alcoholic drink, with characteristic colour (red, white, rose), aroma (characteristic of the grape variety that gave rise to the juice) and taste [1]. Grapes are one of the earliest fruits used in the human diet and grape juices have become an economical alternative for grape producing regions [2]. Grape juice can be produced from any grape variety once it has attained appropriate maturity. In traditional wine producing regions, grape juices are produced from *Vitis vinifera* grape varieties, while, in Brazil, the grape juices are essentially produced from *Vitis labrusca* grape varieties that present as the main feature the preservation of the natural flavour after pasteurisation. Most *Vitis vinifera* grape varieties have an unpleasant taste after the heat treatment, while the American grape varieties keep in the juice the characteristic aroma of the natural grape [3]. Due to consumers' preferences for aroma, colour and flavour, grape juice is mainly made from American cultivars of *Vitis labrusca* species. However, cultivars of *Vitis vinifera* species, although being preferentially used for the wine production, are also used for the juice production [4], especially the 'Cabernet Sauvignon' cultivar [5]. Grape juices have been broadly studied due to their composition in phenolic compounds and their potential beneficial effects on human health, specifically the ability to prevent various diseases associated with oxidative stress, including cancers, cardiovascular and neurodegenerative diseases [6–9]. Grape juices have a high level of phenolic compounds responsible for functional properties; however, the phenolic compounds present in grape juice may also contribute to defining sensory characteristics of this product. A market survey in USA, in 2007, revealed that sixty percent of consumers who purchased functional beverages chose them in part for the presence of antioxidants [10].

2. Grape Juice Production and Phenolic Composition

Grape juice world production is estimated to be between 11 and 12 million hectolitres, where the main producing and consuming countries of this beverage are the United States of America, Brazil and Spain [11]. In many European countries, grape juice is produced from *Vitis vinifera* grape varieties [12]. However, in the United States, the main cultivars used for juice production are mainly Concord and Muscadine (*Vitis rotundifolia*) cultivars [13,14]. Juices from Brazil are produced from American and hybrid grape varieties, the *Vitis labrusca* cultivars "Isabel", "Bordô" and "Concord" being the base for Brazilian grape juices [15]. Among the grape varieties of *Vitis labrusca*, the Concord variety is the most used for the production of grape juice, due to its quality of producing a very aromatic juice with good nutritional properties and being well accepted by the consumers [16]. The "Isabel Precoce" (*Vitis labrusca*) originating from a spontaneous somatic mutation of the "Isabel" cultivar presents good productivity, early maturation and the same characteristics of the original cultivar [17,18]. On the other hand, hybrid cultivars "BRS Cora" and "BRS Violeta" are used for colour improvement in juices deficient in this sensory attribute, where it is recommended to mix in a proportion of 15 to 20% of the juice formulation [19,20].

Different processing technologies are available at an industrial scale for grape juice production, the main techniques being the "Hot press" (HP), "Cold press" (CP) and "Hot Break" (HB) processes [13,21]. These processes have undergone continuous improvements to increase the quality and yield of the grape juice production. In the HP process, grapes are destemmed, crushed and heated to temperatures ranging from 60 to 62 °C to facilitate the extraction of the substances inside the grape cells (Figure 1). Pectinase enzymes are added to degrade pectins and facilitate juice separation. The heated grapes are deposited in stainless steel tanks equipped with internal stirring to facilitate the extraction of the compounds contained in the grape skins, a stage known as maceration. Maceration time ranges from 30 to 90 min, depending on the grape variety, agitation intensity, temperature and colour intensity desired [22]. The difference between the "Hot press" and the "Cold press" processes are small, the main difference being the performance of the maceration at room temperature and the addition of sulphur dioxide (SO_2) after grape crushing to inhibit the action of oxidative enzymes and undesirable microorganisms. Enzymatic preparations based on pectinases are also added in order to degrade the grape skin structures and facilitate the release of phenolic compounds in the juice. Lastly, in the "Hot Break" juice processing, the grapes are crushed and heated at temperatures higher than 75 °C, usually between 77 °C and 82 °C, for a short time to deactivate the polyphenoloxidases, and then cooled to 60 °C to add pectinase enzyme [1,14]. To enhance the juice phenolic and aromatic composition, maceration during juice extraction is performed [14]. In addition, heating of the crushed grape has the main objective to facilitate the release of the juice and anthocyanins responsible for the juice colour. After grape pressing, the cloudy juice is subjected to clarification treatments to remove the suspended solids, usually with the use of rotary vacuum filter or industrial centrifuges, following stabilisation, pasteurisation at 85 °C for three minutes and hot bottling [13,21].

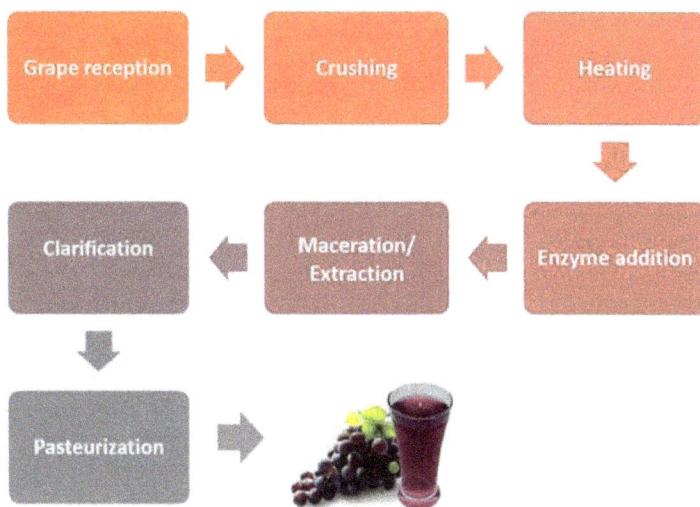

Figure 1. Simplified grape juice flow diagram of hot pressing.

From these technological processes, the continuous hot-pressing process is the most adopted technique worldwide since the cold press process extracts a very low juice yield (~18%) [22,23]. It is also important to consider, in the choice of the grape juice processing technology, the efficient extraction of phenolic compounds from skins, since these compounds are of extreme importance to guarantee a high quality level of the final product, namely its colour and antioxidant capacity but also juice astringency and bitterness [24]. Therefore, the extraction process is essential for the red grape juices' chemical composition and sensory attributes.

In the grape berries, the phenolic compounds are distributed in the different parts of the fruit, as shown in Figure 2. Consequently, the phenolic compounds present in grape juice are mainly those ones extracted from the grape skins and, to a lesser extent, those extracted from the grape seeds [21]. Pastrana-Bonilla et al. [25] found a total concentration of phenolic compounds of about 2178.8, 374.6 and 23.8 mg/g GAE (gallic acid equivalent) in skin, seed and pulp, respectively. Grape skins are the main source of grape phenolic compounds changing their content with grape variety, soil composition, climate, geographic origin, cultivation practices or exposure to diseases, such as fungal infections and reactions occurring during storage [1,26–28]. The total phenolic compounds content in grapes juices (400 to 3000 mg/L) depends on the grape variety, grape maturity, geographical origin and soil type, sunlight exposure, and many other factors [29], besides the grape juice processing technology, such as grape juice extraction, contact time between juice and the grape solid parts (skins and seeds), pressing, thermal and enzymatic treatments. Addition of sulphur dioxide and tartaric acid also interferes with the quantity and the nature of the phenolic compounds present in grape juice [30]. The thermal treatment of intact or crushed grapes enhances the release of phenol compounds, as a consequence of both the increase mass transfer [31] and the higher solubility of cell components [32]. Normally, the temperatures used in the extraction process are not lower than 60 °C [33,34] for different times according to the processing technology. As referred by Celotti and Rebecca [35], the combinations of time/temperature influenced in a different way the extraction yield of the phenolic compounds, according to the molecular type; therefore, at 55 °C, tannin extraction is favoured over red pigment extraction, but, at 63 °C, the maximum extraction of anthocyanins (red pigments) happens after 20 min. Given the importance of anthocyanin extraction during grape juice production, the adequate combination of skin contact time and temperature treatment during maceration is essential to achieving grape juice consumer's acceptable chromatic characteristics as well as antioxidant activity.

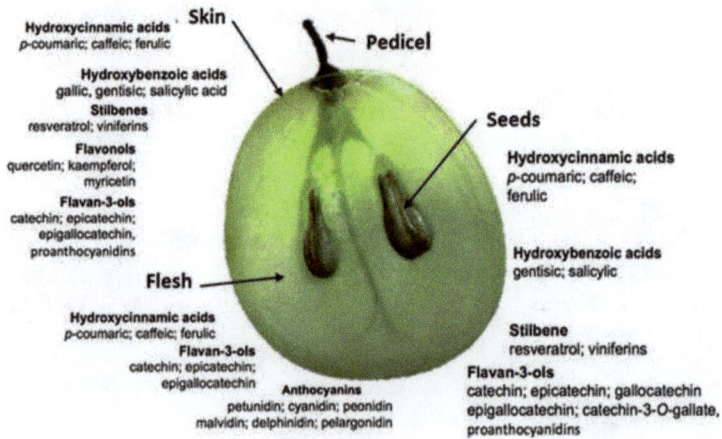

Figure 2. Schematic structure of a ripe grape berry and phenolic pattern biosynthesis distribution between several organs and tissues (indicated by arrows).

Total phenolic compounds according to their chemical structure are classified into flavonoid and non-flavonoid compounds.

Flavonoids are found mainly in grape seeds and skins. Proanthocyanidins in grapes are present mainly in the berry skin and seed. Grape seed proanthocyanidins comprise only (+)-catechin, (−)-epicatechin and procyanidins (Figure 3) [36], whereas grape skin proanthocyanidins comprise both prodelphinidins and procyanidins [37,38]. Procyanidins are dimers resulting from the union of monomeric units of flavanols [(+)-catechin, (−)-epicatechin] by C_4–C_8 (procyanidin B1 to B4) or C_4–C_6 (procyanidin B5 to B8) interflavane linkage. Among grape varieties, there are differences in procyanidins concentrations, but their profile remains unchanged among grape varieties; procyanidin B1 is usually more abundant in the skin while B2 is more abundant in seeds [39]. Prodelphinidins are only present in grape skin and their monomers are [(+)-catechin, (−)-epicatechin, (+)-gallocatechin and (−)-epigallocatechin units] (Figure 3). Proanthocyanidins (procyanidins and prodelphinidins) are the major phenolic compounds in grape seed and skins [39], about 60–70% of total polyphenols are stored in grape seeds [40–42]. According to several published works [43,44], on average and on the basis of fresh weight, the concentrations of proanthocyanidins are as follows: total monomers ((+)-catechin and (−)-epicatechin), 2–12 mg/g in seeds and 0.1–0.7 mg/g in skins; total oligomers, 19–43 mg/g in seeds and 0.8–3.5 mg/g in skins and total polymers, 45–78 mg/g in seeds and 2–21 mg/g in skins.

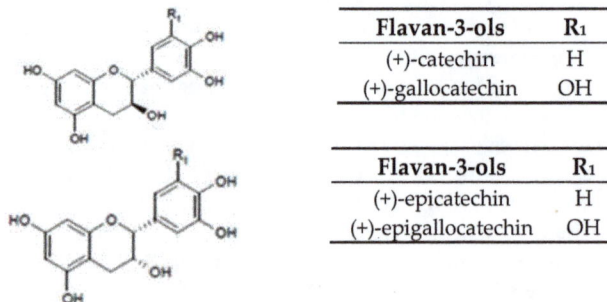

Flavan-3-ols	R_1
(+)-catechin	H
(+)-gallocatechin	OH

Flavan-3-ols	R_1
(+)-epicatechin	H
(+)-epigallocatechin	OH

Figure 3. *Cont.*

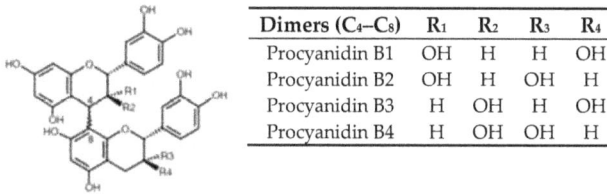

Dimers (C₄–C₈)	R₁	R₂	R₃	R₄
Procyanidin B1	OH	H	H	OH
Procyanidin B2	OH	H	OH	H
Procyanidin B3	H	OH	H	OH
Procyanidin B4	H	OH	OH	H

Figure 3. Monomeric flavanols and procyanidins structures found in grapes.

In grape skins, it is important to notice that each grape species and variety has its unique set of anthocyanins. *Vitis vinifera* grape varieties have only one molecule of glucose at the carbon 3 position forming the 3-*O*-monoglycoside anthocyanins. The glucose part of the anthocyanins can both be unsubstituted or acylated as esters of acetic acid, *p*-coumaric acid, or caffeic acid, Figure 4 (Mazza, 1995). In the species *Vitis labrusca*, *Vitis riparia* and *Vitis rupestrise* hybrids, the glucose molecule appears in positions 3 and 5 of the carbon, forming the 3,5-*O*-diglucoside anthocyanins, this being an important factor in the differentiation of grapes [45,46]. Grape anthocyanins are flavonoids, and the monomeric anthocyanins are six in total: cyanidin, peonidin, pelargonidin, delphinidin, petunidin and malvidin, the latter being the main found in red grape juices [47]. Anthocyanins are the phenolic compounds responsible for the red colour of grape juices [48,49]. Anthocyanin biosynthesis is influenced by several factors such as climatic conditions, temperature, light and cultural practices [38]. The total anthocyanin content in the grape skins from nine grape varieties studied by Jin et al. [50] ranged from 1500 to 30,000 mg malvidin equivalents/kg dry weight of grape skin. However, the grape juice anthocyanins concentration depends on the factors related to the raw material, but the processing technology also exerts a significant influence on the anthocyanin concentration of the grape juice, where the use of heat treatments is fundamental for a greater extraction of the anthocyanins from the grape skins [51].

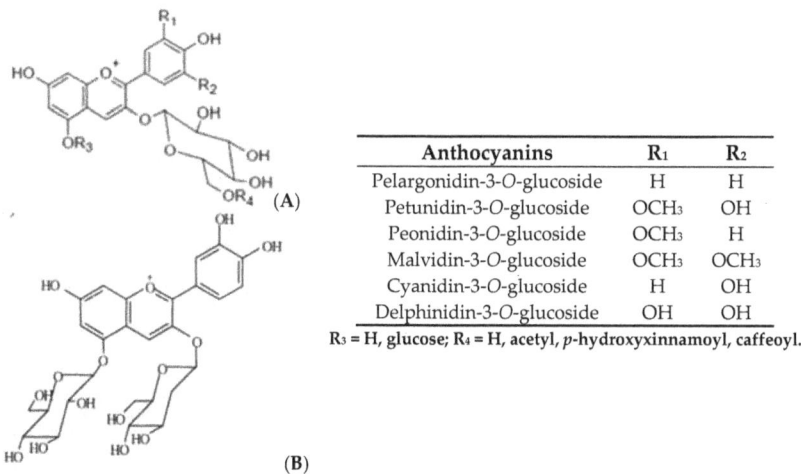

Anthocyanins	R₁	R₂
Pelargonidin-3-*O*-glucoside	H	H
Petunidin-3-*O*-glucoside	OCH₃	OH
Peonidin-3-*O*-glucoside	OCH₃	H
Malvidin-3-*O*-glucoside	OCH₃	OCH₃
Cyanidin-3-*O*-glucoside	H	OH
Delphinidin-3-*O*-glucoside	OH	OH

R₃ = H, glucose; R₄ = H, acetyl, *p*-hydroxyxinnamoyl, caffeoyl.

Figure 4. Anthocyanins structure found in *Vitis vinifera* (**A**) and in *Vitis labrusca* (**B**) red grape varieties.

Thus, anthocyanins are the main phenolic compounds in red grapes juices, while flavan-3-ols are more abundant in white grape juices [52,53].

The flavonols present in the grapes are mainly represented by kaempferol, quercetin and myricetin, and by simple *O*-methylated forms such as isoramnetin [54]. Phenolic acids are divided into benzoic and cinnamic acids. In grapes, phenolic acids are mainly hydroxycinnamic acids found in the skins and pulps, in the form of tartaric esters [45]. The most important benzoic acid are vanillic, syringic and salicylic acids, which appear to be bound to the cell walls and, in particular, gallic acid, which is in the form of the ester of flavanols. Other benzoic acids that are present in lesser amounts are the protocatechic and *p*-hydroxybenzoic acids. The most important cinnamic acids are ferrulic, *p*-coumaric and caffeic acids [45]. Grapes also contain C_6–C_3–C_6 stilbenes, such as *trans*-resveratrol, *cis*-resveratrol, and *trans*-resveratrol glucoside [47]. Since the 1990s, resveratrol has been extensively studied in grapes and their derivatives because of its bioactive activities, such as antioxidant, anti-inflammatory, antimicrobial, anticancer, antiaging, cardioprotective effects and inhibition of platelet aggregation [55,56]. Grape juice is considered a good source of resveratrol; however, the concentration of resveratrol in grapes and consequently in the grape juices depend on the climatic conditions, grape variety and grape growing conditions as well as on the juice processing method used. According to Sautter et al. [57], the concentration of *trans*-resveratrol in grape juice ranged from 0.19 to 0.90 mg/L.

3. Biological Activity of Phenolic Compounds Present in Grape Juices

Phenolic compounds are secondary metabolites, primarily located in the epidermal layer of grape berry skin and seeds and are known as important bioactive compounds, as well as major contributors of the biological activities in products derived from grapes, such as grape juice. Therefore, the evaluation of the beneficial chemical compounds and bioactivity of commercial grape juices is essential to consumer's knowledge in order to provide information about their possible health benefits and bioactivity. Many of these benefits are linked to the antioxidant compounds found in grape juice, such as flavonoids (anthocyanins, proanthocyanidins), phenolic acids and resveratrol among others [5,58]. Therefore, grape juice is consumed not only due to its appreciated sensory characteristics but also because it is a cheap source of phenolic compounds that exert beneficial health effects when consumed [59–61], as shown in Table 1.

The antioxidant activity of the phenolic compounds depends on their structure, particularly on the number and position of the hydroxyl groups and the nature of the substitutions on the aromatic rings [62]. Quercetin is considered to be one of the phenolic compounds with the highest antioxidant activity. Among the different flavonols, it is possible to establish the following decreasing order of antioxidant activity, quercetin, myricetin and kaempferol, which differ concerning their hydroxyl substitution pattern on the B ring. It is noted that the presence of the third hydroxyl group on aromatic B ring, at the C-5′ position, did not result in antioxidant capacity of myricetin higher than quercetin. Catechin, which presented the same number of hydroxyl groups in the molecule as quercetin, presented significantly lower antioxidant activity. This is due to the fact that the structure of catechin does not have unsaturated bonds at the C2–C3 position in conjunction with the oxo (−C=O) on C ring, which comparatively gives quercetin a higher antioxidant activity. With the addition of a hydroxyl group on the catechin molecule B ring, this compound is called epigallocatechin and, with this new structure, there is an increase in the antioxidant activity, but not equivalent to quercetin [63]. In general, aglycones are more potent antioxidants than their corresponding glycosides [64]. In relation to the phenolic acids, it is possible to observe that hydroxycinnamic acids are more effective antioxidants than hydroxybenzoic acids. This is due to the conjugation through the double bonds of the ring to the −CH=CH-COOH of the cinnamic acid structure, which enhances the ability to stabilize free radicals. However, it must be emphasised that gallic acid has more antioxidant activity than catechin, which has five hydroxyl groups in its structure [63].

Table 1. Phenolic compounds (mg/L) from grape juice obtained in different conditions of maceration, from different grape varieties and from organic and conventional grape production [22,60,61].

Phenolic Compounds	Temperature/Enzyme		Grape Variety			Grape Production	
	50 °C/3.0 g	60 °C/3.0 g	Isabel Precoce	BRS Cora	BRS Violeta	Organic	Conventional
(+)-Catechin	9.9 ± 0.5	11.2 ± 3.4	4.7 ± 0.1	12.4 ± 0.3	19.8 ± 0.4	500.52 ± 12.33	79.89 ± 30.19
(−)-Epicatechin	0.8 ± 0.7	1.7 ± 0.3	1.0 ± 1.0	1.4 ± 0.5	0.6 ± 0.1	53.48 ± 19.78	14.40 ± 0.77
(−)-Epicatechin gallate	2.2 ± 1.2	0.8 ± 0.1	1.6 ± 0.1	1.2 ± 0.0	1.9 ± 0.2	-	-
(−)-Epigallocatechin gallate	2.1 ± 0.5	1.8 ± 0.3	-	-	-	-	-
(−)-Epigallocatechin	-	-	0.9 ± 0.1	4.7 ± 0.4	6.2 ± 0.1	-	-
Procyanidin A2	2.6 ± 0.2	3.0 ± 0.2	2.8 ± 0.2	2.9 ± 0.2	3.6 ± 0.1	-	-
Procyanidin B1	34.4 ± 2.4	36.8 ± 0.8	47.1 ± 0.1	37.2 ± 0.6	44.2 ± 0.3	-	-
Procyanidin B2	13.1 ± 3.1	16.1 ± 2.5	14.3 ± 0.1	16.3 ± 0.7	17.5 ± 0.5	-	-
trans-Resveratrol	0.90 ± 0.3	0.90 ± 0.4	-	-	-	3.73 ± 0.19	2.24 ± 0.07
Malvidin 3,5-diglucoside	5.2 ± 1.0	6.4 ± 0.1	1.8 ± 0.0	0.7 ± 0.0	11.7 ± 0.0	721.26 ± 20.99	189.43 ± 1.29
Malvidin 3-glucoside	8.1 ± 3.6	11.2 ± 0.5	0.9 ± 0.1	-	1.6 ± 0.2	23.91 ± 2.59	47.42 ± 0.73
Cyanidin-3,5-diglucoside	4.8 ± 1.0	5.6 ± 1.0	-	11.8 ± 0.1	38.0 ± 0.6	785.53 ± 39.56	152.02 ± 6.98
Cyanidin 3-glucoside	5.2 ± 3.2	8.0 ± 0.1	3.0 ± 0.0	1.4 ± 0.1	32.7 ± 0.5	21.72 ± 4.17	7.17 ± 0.59
Delphinidin 3-glucoside	3.3 ± 2.8	6.0 ± 1.2	-	11.7 ± 0.2	73.7 ± 1.2	17.79 ± 1.01	12.15 ± 0.09
Peonidin 3-glucoside	2.2 ± 1.0	3.0 ± 0.1	0.2 ± 0.1	0.3 ± 0.1	0.2 ± 0.0	2.45 ± 0.66	10.84 ± 0.11
Pelargonidin 3-glucoside	2.4 ± 1.4	3.6 ± 0.4	-	6.7 ± 0.1	6.7 ± 0.1	-	-
Gallic acid	2.3 ± 0.4	2.8 ± 0.5	1.8 ± 0.1	13.6 ± 0.1	10.5 ± 0.8	16.96 ± 0.39	11.51 ± 0.10
Caffeic acid	15.7 ± 1.0	17.2 ± 1.6	8.6 ± 0.1	35.8 ± 0.5	28.9 ± 0.4	29.95 ± 1.57	14.08 ± 0.17
Cinnamic acid	1.2 ± 0.7	2.0 ± 0.2	0.5 ± 0.0	0.6 ± 0.2	1.9 ± 0.1	-	-
Chlorogenic acid	2.3 ± 0.1	2.5 ± 0.3	4.1 ± 0.1	8.3 ± 0.3	21.3 ± 0.6	-	-
p-Coumaric acid	1.4 ± 0.0	1.7 ± 0.1	2.6 ± 0.1	4.5 ± 0.4	9.0 ± 0.1	11.23 ± 0.16	10.73 ± 0.51
Kaempferol	0.9 ± 0.2	1.2 ± 0.2	-	-	-	2.67 ± 0.02	3.01 ± 0.67
Myricetin	0.2 ± 0.1	0.3 ± 0.1	-	-	-	7.99 ± 0.99	6.98 ± 0.90
Isorhamnetin	1.2 ± 0.6	1.6 ± 0.2	-	-	-	-	-
Rutin	1.5 ± 0.1	1.8 ± 0.3	-	-	-	-	-
Quercetin	0.10 ± 0.00	0.17 ± 0.06	-	-	-	3.91 ± 0.08	4.27 ± 0.54
DPPH (TE mmol/L)	-	-	-	-	-	54.19 ± 0.24	40.76 ± 0.71
ABTS (TE mmol/L)	-	-	-	-	-	51.90 ± 0.33	31.09 ± 0.17

DPPH—2,2-diphenyl-1-picrylhydrazyl; ABTS—2,2′-Azino-bis (3-ethylbenzothiazoline-6-sulfonic acid).

Grapes are one of the richest sources of polyphenols among fruits [25], being rich in a wide range of phenolic compounds, many of them renowned for their therapeutic or health promoting properties [65,66]. The bioactive compounds from grapes juices mainly include simple phenolic, flavonoids (anthocyanins, flavanols, flavonols), stilbenes (resveratrol) and phenolic acids, shown to have benefits related to human health by conferring the ability to sequester reactive oxygen species (ROS), such as hydroxyl radical and singlet oxygen [67]. Several clinical studies on grapes and their derivatives have demonstrated these properties, including protection against cardiovascular diseases [68,69], atherosclerosis [70], hypertension [71], cancer [72], diabetes [73] and neurological problems [74]. The mechanism of action has been attributed to antioxidant activity [75], lipid regulation [76,77], anti-inflammatory effects [78], anti-cancer, antimicrobial, antiviral, cardio protective, neuroprotective, and hepatoprotective activities [1,40,41,48,79–83].

In general, grapes produced under organic farming system are increasing around the world, since they are perceived by the public as safer, and healthier, than those produced by conventional agriculture, as they do not use chemical pesticides and fertilizers for growing. These grapes are more susceptible to the action of phytopathogens, inducing the syntheses of higher amounts of phenolic compounds as protection and defence [84]. Dani et al. [59] observed that the choice of agricultural practice (organic versus conventional) resulted in different amounts of resveratrol, anthocyanins, and tannins in grape juices. This difference is due to the fact that no pesticides are used in organic vineyards and by the fact that usually they have a longer ripening period than conventional vineyards, and as flavonoids are formed during this period, it is believed that organic vineyards yield grapes with a higher phenolic content [85,86].

4. Sensory Characteristics of Grape Juices

During grape juice processing, heat treatments can adversely impact grape juice flavour [87]. Other physical treatments, applied to grape juices, can also interfere with their sensory quality. According to Treptow et al. [88], the storage period and irradiation promoted few physicochemical changes in juices from 'Niagara Branca' and 'Trebbiano' grapes, whereas, sensorially, irradiation reduced the intensity of flavour and colour attributes for both cultivars. While in juices made from white grapes, UV-C irradiation does not improve juice quality, in juices from red grapes UV-C irradiation influenced the physicochemical parameters of 'Isabel' cultivar. Good results were obtained with dosages of 2 kJ m^{-2} (improved of aroma and colour intensities). According to these authors [88], UV-C irradiation can keep the juices microbiologically stable, without spoiling their sensory quality.

Grape juices are known to have high concentration of phenolic compounds that contribute to define sensory characteristics of grape juice [89], namely their colour, taste and flavour.

Colour is the most important attribute used, along with other variables, as an indicator of the grape juice quality observed by the consumers. This characteristic is directly dependent on the juice phenolic composition, namely on the anthocyanins present in the grape skin. Anthocyanins participate in many reactions that promote changes in grape juice colour, mainly through copigmentation and formation of polymeric pigments [90]. Hue and colour intensity can provide information with respect to possible defects or quality of the raw material [91]. The colour of grape juices can vary according to the origin and region of the grapes, juice processing technology [92] and physical and chemical characteristics of pigments (anthocyanins) that are present in the juice [93]. Concentrating a grape-juice, by reverse osmosis, may lead to an increase in total acidity, colour intensity, anthocyanin and phenolic compounds, proportional to the volumetric concentration factor. The increase in soluble solids may be associated with browning of the concentrated juice [91]. Moreover, and according to Gurak et al. [91], as a consequence of the buffer characteristics of fruit juices, the stability of anthocyanins was favoured by the low pH, lack of vitamin C and high concentration of sugar and the pH of the concentrated juice, compared to single-strength juice. In a work performed by Meullenet et al. [94], to compare the use of an internal density plot method to the external response surface approach by optimizing the formulation of muscadine grape juices, the authors found that a red juice appearance is, on average, preferable by consumers to white juices.

Huckleberry and co-workers [95] studied 14 traditional wine grapes grown in Arkansas, aiming to find their suitability for varietal grape juice. Chemical and sensory analyses were performed. The preferred treatments for white grape juices were the immediate pressing treatment of Niagara and Aurore and the 24-h skin contact treatment of Niagara, Verdelet and Vidal. A trained sensory panel ranked the flavour of Verdelet and Aurore as high as the Niagara. In general, a non-heat treatment was ranked highest for flavour of the red juices, with the exception of Gewürztraminer (a pink juice when heated), while white grape varieties were closely ranked for colour (Vidal and Chardonnay grape juices were ranked lowest after storage). For red and pink juices, heat treatments were ranked higher than non-heat treatments in terms of colour attribute [95].

Grape juice consists of 81 to 86% of water and a high concentration of sugars (glucose and fructose). It may also present high acidity due to the existence of tartaric, malic and citric acids. These acids guarantee a low pH value, assuring the equilibrium between sour and sweet tastes [91]. Among the bioactive compounds present, phenolic compounds are of great importance because their characteristics are directly or indirectly related to the juice quality and affect its colour and astringency [96].

Meullenet et al. [94] in Muscadine grape juices found that the attribute "Musty" was also correlated with the perception of Muscadine flavour. Juices perceived as having a general "grape flavour" were also found to be high in metallics. Green/unripe flavour was associated with sour and astringent juices. Products perceived as high in sweetness were also high in floral and apple/pear flavours.

A typically sour grape juice is made from unripe grapes. They can be processed into two products: verjuice and sour grape sauce. In the "Middle Ages", verjuice was widely used all over Western Europe. The verjuice ("verjus" or "agraz") is obtained by pressing unripe grapes, while sour grape sauce is derived from verjuice that undergoes an additional concentration step followed by salting [97]. In Tuscany (Italy), there is a similar product called, "agresto" [98]. In Iran and Turkey, it is called "abe ghureh" (Persian) and "koruk suyu" (Turkish), respectively [99]. In Iran, it is used in salad dressings and digestive drinks [100]. Recently, unripe grape juice has been used as an alternative to vinegar or lemon juice [97,100]. Unripe grape juice can also be used as a food preservative due to its high organic acid content [101] and high concentration of phenolic compounds [102].

Unripe grape juice is characterised by high acidity, low sugar content and a sour/tart taste. It also presents a high content of phenolic compounds, the latter of which have an astringent character [103]. The concentration of most polyphenolic compounds in general, and astringent tannins in particular, reaches a maximum of around 45 days after flowering [45], a time approximately corresponding to the harvest date for producing unripe grape juice.

In the sensory evaluation of verjuice done by Matos et al. [104], the most frequently used descriptors perceived for taste were "acid" (74.19%), "astringent" (51.61%), "salty" (35.48%) and "sweet" (25.80%), while the most common aroma terms were "herbaceous" (50.00%), "cooked apple" (43.55%), "pear" (29.03%), "floral" (29.03%) and "green apple" (25.81%) (Figure 5).

Figure 5. To make verjuice, sour grapes, very small and very sour, are used (**A**); a grape vinegar is made in Lebanon with sour grapes (**B**) in order to have something acidic to add to salads and stuffed vegetables. This vinegar is called "husrum" or verjuice. The grapes are crushed and the juice collected; the sour juice is salted and simmer over medium heat, skimming the foam until it disappears. After cooling, it is poured into a glass or clay bottle, adding a thin layer of olive oil on top to prevent microbial contamination and oxidation. The traditional recipe retrieved from: [105]

5. Conclusions

The unique combination of grape phenolic compounds, including flavonoids (anthocyanins, proanthocyanins), and stilbenes, makes grape, and grape juices, a promising source for the development of novel nutraceutical products.

The grape juices' high concentration of phenolic compounds contributes to defining their sensory characteristics in terms of colour, taste and flavour.

In the last few years, there has been a wide range of food additives and nutritional products originating from grapes, distributed in the worldwide market; however, we must retain that the consumers search not only the nutraceutical products characteristics, but also their palatability.

Conflicts of Interest: The authors declare no conflict of interest.

References

1. Rizzon, L.A.; Meneguzzo, J. *Suco de Uva*, 1st ed.; Embrapa Technological Information: Brasília, Brazil, 2007.
2. Ali, K.; Maltese, F.; Choi, Y.; Verpoorte, R. Metabolic constituents of grapevine and grape-derived products. *Phytochem. Rev.* **2010**, *9*, 357–378. [CrossRef] [PubMed]
3. Rizzon, L.A.; Manfroi, V.; Meneguzo, J. *Elaboração de Suco de uva na Propriedade Vitícola*; Embrapa-CNPUV: Bento Gonçalves, Brazil, 1998; pp. 1–24.
4. Camargo, U.A.; Maia, J.D.G.; Ritschel, P.S. *Novas Cultivares Brasileiras de Uva*; Embrapa Uva e Vinho: Bento Gonçalves, Brazil, 2010; 64p.
5. Dutra, M.C.P.; Lima, M.S.; Barros, A.P.A.; Mascarenhas, R.J.; Lafisca, A. Influência da variedade de uvas nas características analíticas e aceitação sensorial do suco artesanal. *Rev. Bras. Prod. Agroind.* **2014**, *16*, 265–272. [CrossRef]
6. Soundararajan, R.; Wishart, A.D.; Rupasinghe, H.P.V.; Arcellana-Panlilio, M.; Nelson, C.M.; Mayne, M.; Robertson, G.S. Quercetin-3-glucoside protects neuroblastoma (SH-SY5Y) cells in vitro against oxidative damage by inducing SREBP-2 mediated cholesterol biosynthesis. *J. Biol. Chem.* **2008**, *283*, 2231–2245. [CrossRef] [PubMed]
7. Gollücke, A.P.B.; Catharino, R.R.; Souza, J.C.; Eberlin, M.N.; Tavares, D.Q. Evolution of major phenolic components and radical scavenging activity of grape juices through concentration process and storage. *Food Chem.* **2009**, *112*, 868–873. [CrossRef]
8. Fraga, C.G.; Galleano, M.; Verstraeten, S.V.; Oteiza, P.I. Basic biochemical mechanisms behind the health benefits of polyphenols. *Mol. Asp. Med.* **2010**, *31*, 435–445. [CrossRef] [PubMed]
9. Yamagata, K.; Tagami, M.; Yamori, Y. Dietary polyphenols regulate endothelial function and prevent cardiovascular disease. *Nutrition* **2015**, *31*, 28–37. [CrossRef] [PubMed]
10. Mintel. *Functional Beverages-US August*; Market Research Report; Mintel International Group Ltd.: Chicago, IL, USA, 2007.
11. OIV—Organisation Internationale de la Vigne et du Vin. *Vine and Wine Outlook*; OIV: Paris, France, 2012; ISBN 979-10-91799-56-0.
12. Soyer, Y.; Koca, N.; Karadeniz, F. Organic acid profile of Turkish white grapes and grape juices. *J. Food Compos. Anal.* **2003**, *16*, 629–636. [CrossRef]
13. Morris, J.R. Factors influencing grape juice quality. *Hort Technol.* **1998**, *8*, 471–478.
14. Iyer, M.M.; Sacks, G.L.; Padilla-Zakour, O.I. Impact of harvesting and processing conditions on green leaf volatile development and phenolics in concord grape juice. *J. Food Sci.* **2010**, *75*, 297–304. [CrossRef] [PubMed]
15. Rizzon, L.A.; Miele, A. Características analíticas e discriminação de suco, néctar e bebida de uva comerciais brasileiros. *Ciênc Tecnol Aliment.* **2012**, *32*, 93–97. [CrossRef]
16. Stalmach, A.; Edwards, C.A.; Wightman, J.D.; Crozier, A. Identification of (Poly)phenolic Compounds in Concord Grape Juice and Their Metabolites in Human Plasma and Urine after Juice Consumption. *J. Agric. Food Chem.* **2011**, *59*, 9512–9522. [CrossRef] [PubMed]
17. Ribeiro, T.P.; Lima, M.A.C.; Alves, R.E. Maturação e qualidade de uvas para suco em condições tropicais, nos primeiros ciclos de produção. *Pesq. Agrop. Bras.* **2012**, *47*, 1057–1065. [CrossRef]
18. Camargo, U.A. 'Isabel Precoce': Alternativa Para a Vitivinicultura Brasileira. In *Comunicado Técnico N° 54*; Embrapa Uva e Vinho: Bento Gonçalves, Brazil, 2004.

19. Camargo, U.A.; Maia, J.D.G. "BRS Cora" nova cultivar de uva para suco, adaptada a climas tropicais. In *Comunicado Técnico N° 53*; Embrapa Uva e Vinho: Bento Gonçalves, Brazil, 2004.

20. Camargo, U.A.; Maia, J.D.G.; Nachtigal, J.C. "BRS Violeta" nova cultivar de uva para suco e vinho de mesa. In *Comunicado Técnico N° 63*; Embrapa Uva e Vinho: Bento Gonçalves, Brazil, 2005.

21. Morris, J.R.; Striegler, K.R. *Processing Fruits: Science and Technology*, 2nd ed.; CRC Press: Boca Raton, FL, USA, 2005.

22. Lima, M.S.; Dutra, M.C.P.; Toaldo, I.M.; Corrêa, L.C.; Pereira, G.L.; Oliveira, D.; Bordignon-Luiz, M.T.; Ninowd, J.L. Phenolic compounds, organic acids and antioxidant activity of grape juices produced in industrial scale by different processes of maceration. *Food Chem.* **2015**, *188*, 384–392. [CrossRef] [PubMed]

23. Monrad, J.K.; Suárez, M.; Motilva, M.J.; King, J.W.; Srinivas, K.; Howard, L.R. Extraction of anthocyanins and flavan-3-ols from red grape pomace continuously by coupling hot water extraction with a modified expeller. *Food Res. Int.* **2014**, *65*, 77–87. [CrossRef]

24. Kelebek, H.; Canbas, A.; Selli, S. Effects of different maceration times and pectolytic enzyme addition on the anthocyanin composition of *Vitis vinifera* cv. Kalecik karasi wines. *J. Food Process. Preserv.* **2009**, *33*, 296–311. [CrossRef]

25. Pastrana-Bonilla, E.; Akoh, C.C.; Sellappan, S.; Krewer, G. Phenolic content and antioxidant capacity of Muscadine grapes. *J. Agric. Food Chem.* **2003**, *51*, 5497–5503. [CrossRef] [PubMed]

26. Bruno, G.; Sparapano, L. Effects of three esca-associated fungi on *Vitis vinifera* L. Changes in the chemical and biological profile of xylem sap from diseased cv. Sangiovese vines. *Physiol. Mol. Plant Pathol.* **2007**, *71*, 210–229. [CrossRef]

27. Rombaldi, C.V.; Bergamasqui, M.; Lucchetta, L.; Zanuzo, M.; Silva, J.A. Vineyard yield and grape quality in two different cultivation systems. *Rev. Bras. Frutic.* **2004**, *26*, 89–91. [CrossRef]

28. Frankel, E.N.; Bosanek, C.A.; Meyer, A.S.; Silliman, K.; Kirk, L.L. Commercial grape Juices inhibit the in vitro oxidation of human low density lipoproteins. *J. Agric. Food Chem.* **1998**, *46*, 834–838. [CrossRef]

29. Garrido, J.; Borges, F. Wine and grape polyphenols—A chemical perspective. *Food Res. Int.* **2013**, *54*, 1844–1858. [CrossRef]

30. Shahidi, F.; Naczk, M. Phenolic compounds in fruits and vegetables. In *Phenolics in Food and Nutraceuticals*; CRC Press: New York, NY, USA, 2004; pp. 144–151.

31. Corrales, M.; García, A.F.; Butz, P.; Tauscher, B. Extraction of anthocyanins from grape skins assisted by high hydrostatic pressure. *J. Food Eng.* **2009**, *90*, 415–421. [CrossRef]

32. El Darra, N.E.; Grimi, N.; Maroun, R.G.; Louka, N.; Vorobiev, E. Pulsed electric field, ultrasound, and thermal pretreatments for better phenolic extraction during red fermentation. *Eur. Food Res. Technol.* **2013**, *236*, 47–56. [CrossRef]

33. El Darra, N.E.; Grimi, N.; Vorobiev, E.; Louka, N.; Maroun, R. Extraction of polyphenols from red grape pomace assisted by pulsed ohmic heating. *Food Bioprocess Technol.* **2013**, *6*, 1281–1289. [CrossRef]

34. Sacchi, K.L.; Bisson, L.F.; Adams, D.O. A review of the effect of winemaking techniques on phenolic extraction in red wines. *Am. J. Enol. Vitic.* **2005**, *56*, 197–206.

35. Celotti, E.; Rebecca, S. Recenti esperienze di termomacerazione delle uve rosse. *Industrie delle Bevande* **1998**, *155*, 245–255.

36. Prieur, C.; Rigaud, J.; Cheynier, V.; Moutounet, M. Oligomeric and polymeric procyanidins from grape seeds. *Phytochemistry* **1994**, *3*, 781–784. [CrossRef]

37. Souquet, J.M.; Cheynier, V.; Brossaud, F.; Moutounet, M. Polymeric proanthocyanidins from grape skins. *Phytochemistry* **1996**, *43*, 509–512. [CrossRef]

38. He, F.; Mu, L.; Yan, G.L.; Liang, N.N.; Pan, Q.H.; Wang, J.; Reeves, M.J.; Duan, C.Q. Biosynthesis of Anthocyanins and Their Regulation in Colored Grapes. *Molecules* **2010**, *15*, 9057–9091. [CrossRef] [PubMed]

39. Fuleki, T.; Ricardo-Da-Silva, J.M. Effects of cultivar and processing method on the contents of catechins and procyanidins in grape juice. *J. Agric. Food Chem.* **2003**, *51*, 640–646. [CrossRef] [PubMed]

40. Shrikhande, A.J. Wine by-products with health benefits. *Food Res. Int.* **2000**, *33*, 469–474. [CrossRef]

41. Wroblewski, K.; Muhandiram, R.; Chakrabartty, A.; Bennick, A. The molecular interaction of human salivary histatins with polyphenolic compounds. *Eur. J. Biochem.* **2001**, *268*, 4384–4397. [CrossRef] [PubMed]

42. Auger, C.; Teissedre, P.L.; Gerain, P.; Lequeux, N.; Bornet, A.; Serisier, S.; Besançon, P.; Caporiccio, B.; Cristol, J.P.; Rouanet, J.M. Dietary wine phenolics catechin, quercetin, and resveratrol efficiently protect hypercholesterolemic hamsters against aortic fatty streak accumulation. *J. Agric. Food Chem.* **2005**, *53*, 2015–2021. [CrossRef] [PubMed]

43. Jordão, A.M.; Ricardo-da-Silva, J.M.; Laureano, O. Evolution of catechins and oligomeric procyanidins during grape maturation of Castelão Francês and Touriga Francesa. *Am. J. Enol. Vitic.* **2001**, *52*, 230–234.

44. Sun, B.S.; Ricardo-da-Silva, J.M.; Spranger, M.I. Quantification of catechins and proanthocyanidins in several Portuguese grapevine varieties and red wines. *Ciência Téc Vitiv* **2001**, *16*, 23–34.

45. Ribéreau-Gayon, P.; Dubourdieu, D.; Donèche, B.; Lonvaud, A. *Trattato di Enologia: Microbiologia del vino e Vinificazioni*, 2nd ed.; Edagricole: Bologna, Italy, 2005.

46. Rubilar, M.; Pinelo, M.; Shene, C.; Sineiro, J.; Nunez, M.J. Separation and HPLC-MS identification of phenolic antioxidants from agricultural residues: Almond hulls and grape pomace. *J. Agric. Food Chem.* **2007**, *55*, 10101–10109. [CrossRef] [PubMed]

47. Jackson, R.S. *Wine Science Principle and Applications*, 3rd ed.; Academic Press: Burlington, VT, USA, 2008.

48. Dopico-Garcia, M.S.; Fique, A.; Guerra, L.; Afonso, J.M.; Pereira, O.; Valentao, P.; Andrade, P.B.; Seabra, R.M. Principal components of phenolics to characterize red Vinho Verde grapes: Anthocyanins or non-coloured compounds? *Talanta* **2008**, *75*, 1190–1202. [CrossRef] [PubMed]

49. Vilela, A.; Cosme, F. Drink Red: Phenolic Composition of Red Fruit Juices and Their Sensorial Acceptance. *Beverages* **2016**, *2*, 29. [CrossRef]

50. Jin, Z.M.; He, J.J.; Bi, H.Q.; Cui, X.Y.; Duan, C.Q. Phenolic Compound Profiles in Berry Skins from Nine Red Wine Grape Cultivars in Northwest China. *Molecules* **2009**, *14*, 4922–4935. [CrossRef] [PubMed]

51. Cabrera, S.G.; Jang, J.I.H.; Kim, S.T.; Lee, Y.R.; Lee, H.J.; Chung, H.S.; Moon, K.D. Effects of processing time and temperature on the quality components of Campbell grape juice. *J. Food Process. Preserv.* **2009**, *33*, 347–360. [CrossRef]

52. Bagchi, D.; Bagchi, M.; Stohs, S.J.; Das, D.K.; Ray, C.A.; Kuszynski, S.S.; Joshi, H.G. Free radicals and grape seed proanthocyanidin extract: Importance in human health and disease prevention. *Toxicology* **2000**, *148*, 187–197. [CrossRef]

53. Cantos, E.; Espin, J.C.; Tomas-Barberan, F.A. Varietal differences among the polyphenol profiles of seven table grape cultivars studied by LC-DAD-MS-MS. *J. Agric. Food Chem.* **2002**, *50*, 5691–5696. [CrossRef] [PubMed]

54. Makris, D.P.; Kallithrakab, S.; Kefalasa, K. Flavonols in grapes, grape products and wines: Burden, profile and influential parameters. *J. Food Compos. Anal.* **2006**, *19*, 396–404. [CrossRef]

55. Pace-Asciak, C.R.; Rounova, O.; Hahn, S.E.; Diamandis, E.P.; Goldberg, D.M. Wines and grape juices as modulators of platelet aggregation in healthy human subject. *Clin. Chim. Acta* **1996**, *246*, 163–182. [CrossRef]

56. Daglia, M. Polyphenols as antimicrobial agents. *Curr. Opin. Biotechnol.* **2012**, *23*, 1–8. [CrossRef] [PubMed]

57. Sautter, C.K.; Denardin, S.; Alves, A.O.; Mallmann, C.A.; Penna, N.G.; Hecktheuer, L.H. Determinação de resveratrol em sucos de uva no Brasil. *Ciênc. Tecnol. Aliment.* **2005**, *25*, 437–442. [CrossRef]

58. Capanoglu, E.; Vos, R.C.H.; Hall, R.D.; Boyacioglu, D.; Beekwilder, J. Changes in polyphenol content during production of grape juice concentrate. *Food Chem.* **2013**, *139*, 521–526. [CrossRef] [PubMed]

59. Dani, C.; Oliboni, L.S.; Vanderlinde, R.; Bonatto, D.; Salvador, M.; Henriques, J.A.P. Phenolic content and antioxidant activities of white and purple juices manufactured with organically- or conventionally-produced grapes. *Food Chem. Toxicol.* **2007**, *45*, 2574–2580. [CrossRef] [PubMed]

60. Lima, M.S.; Silani, I.S.V.; Toaldo, I.M.; Corrêa, L.C.; Biasoto, A.C.T.; Pereira, G.E.; Bordignon-Luiz, M.T.; Ninow, J.L. Phenolic compounds, organic acids and antioxidant activity of grape juices produced from new Brazilian varieties planted in the Northeast Region of Brazil. *Food Chem.* **2014**, *161*, 94–103. [CrossRef] [PubMed]

61. Toaldo, I.M.; Cruz, F.A.; Alves, T.L.; Gois, J.S.; Borges, D.L.G.; Cunha, H.P.; Silva, E.L.; Bordignon-Luiz, M.T. Bioactive potential of *Vitis labrusca* L. grape juices from the Southern Region of Brazil: Phenolic and elemental composition and effect on lipid peroxidation in healthy subjects. *Food Chem.* **2015**, *173*, 527–535. [CrossRef] [PubMed]

62. Balasundram, N.; Sundram, K.; Samman, S. Phenolic compounds in plants and agri-industrial by-products: Antioxidant activity, occurrence, and potential uses. *Food Chem.* **2006**, *99*, 191–203. [CrossRef]

63. Rice-Evans, C.A.; Miller, N.J.; Paganga, G. Structure-antioxidant activity relationships of flavonoids and phenolic acids. *Free Radic. Biol. Med.* **1996**, *20*, 933–956. [CrossRef]

64. Hidalgo, M.; Sanchez-Moreno, C.; Pascual-Teresa, S. Flavonoid-flavonoid interaction and its effect on their antioxidant activity. *Food Chem.* **2010**, *121*, 691–696. [CrossRef]

65. Georgiev, V.; Ananga, A.; Tsolova, V. Recent Advances and Uses of Grape Flavonoids as Nutraceuticals. *Nutrients* **2014**, *6*, 391–415. [CrossRef] [PubMed]

66. Wada, M.; Kido, H.; Ohyama, K.; Ichibangas, T.; Kishikaw, N.; Ohba, Y.; Nakashima, M.N.; Kurod, N.; Nakashima, K. Chemiluminescent screening of quenching effects of natural colorants against reactive oxygen species: Evaluation of grape seed, monascus, gardenia and red radish extracts as multi-functional food additives. *Food Chem.* **2007**, *101*, 980–986. [CrossRef]
67. Marinova, E.M.; Yanishlieva, N.V. Antioxidant activity and mechanism of action of some phenolic acids at ambient and high temperatures. *Food Chem.* **2003**, *81*, 189–197. [CrossRef]
68. Coimbra, S.R.; Lage, S.H.; Brandizzi, L.; Yoshida, V.; da Luz, P.L. The action of red wine and purple grape juice on vascular reactivity is independent of plasma lipids in hypercholesterolemic patients. *Braz. J. Med. Biol. Res.* **2005**, *38*, 1339–1347. [CrossRef] [PubMed]
69. Dohadwala, M.M.; Vita, J.A. Grapes and Cardiovascular Disease. *J. Nutr.* **2009**, *139*, 1788–1793. [CrossRef] [PubMed]
70. Aviram, M.; Fuhrman, B. Wine flavonoids protect against LDL oxidation and atherosclerosis. Alcohol and Wine in Health and Disease. *Ann. N. Y. Acad. Sci.* **2002**, *957*, 146–161. [CrossRef] [PubMed]
71. Mudnic, I.; Modun, D.; Rastija, V.; Vukovic, J.; Brizic, I.; Katalinic, V.; Kozina, B.; Medic-Saric, M.; Boban, M. Antioxidative and vasodilatory effects of phenolic acids in wine. *Food Chem.* **2010**, *119*, 1205–1210. [CrossRef]
72. Jang, M.; Cai, L.; Udeani, G.O.; Slowing, K.V.; Thomas, C.F.; Beecher, C.W.W.; Fong, H.H.S.; Farnsworth, N.R.; Kinghorn, A.D.; Mehta, R.G.; et al. Cancer Chemopreventive Activity of Resveratrol, a Natural Product Derived from Grapes. *Science* **1997**, *275*, 218–220. [CrossRef] [PubMed]
73. Zunino, S. Type 2 Diabetes and Glycemic Response to Grapes or Grape Products. *J. Nutr.* **2009**, *139*, 1794–1800. [CrossRef] [PubMed]
74. Joseph, J.A.; Shukitt-Hale, B.; Willis, L.M. Grape Juice, Berries, and Walnuts Affect Brain Aging and Behavior. *J. Nutr.* **2009**, *139*, 1813–1817. [CrossRef] [PubMed]
75. Modun, D.; Music, I.; Vukovic, J.; Brizic, I.; Katalinic, V.; Obad, A.; Palada, I.; Dujic, Z.; Boban, M. The increase in human plasma antioxidant capacity after red wine consumption is due to both plasma urate and wine polyphenols. *Atherosclerosis* **2008**, *197*, 250–256. [CrossRef] [PubMed]
76. Ursini, F.; Sevanian, A. Wine Polyphenols and Optimal Nutrition. *Ann. N. Y. Acad. Sci.* **2002**, *957*, 200–209. [CrossRef] [PubMed]
77. Dávalos, A.; Fernández-Hernando, C.; Cerrato, F.; Martínez-Botas, J.; Gómez-Coronado, D.; Gómez-Cordovés, C.; Lasunción, M.A. Red Grape Juice Polyphenols Alter Cholesterol Homeostasis and Increase LDL-Receptor Activity in Human Cells *In Vitro*. *J. Nutr.* **2006**, *136*, 1766–1773. [CrossRef] [PubMed]
78. Castilla, P.; Echarri, R.; Davalos, A.; Cerrato, F.; Ortega, H.; Teruel, J.L.; Lucas, M.F.; Gomez-Coronado, D.; Ortuno, J.; Lasuncion, M.A. Concentrated red grape juice exerts antioxidant, hypolipidemic, and anti-inflammatory effects in both hemodialysis patients and healthy subjects. *Am. J. Clin. Nutr.* **2006**, *84*, 252–262. [CrossRef] [PubMed]
79. Hernandez-Jimenez, A.; Gomez-Plaza, E.; Martinez-Cutillas, A.; Kennedy, J.A. Grape skin and seed proanthocyanidins from Monastrell x Syrah grapes. *J. Agric. Food Chem.* **2009**, *57*, 10798–10803. [CrossRef] [PubMed]
80. Novaka, I.; Janeiroa, P.; Serugab, M.; Oliveira-Brett, A.M. Ultrasound extracted flavonoids from four varieties of Portuguese red grape skins determined by reverse-phase high-performance liquid chromatography with electrochemical detection. *Anal. Chim. Acta* **2008**, *630*, 107–115. [CrossRef] [PubMed]
81. Spáčil, Z.; Nováková, L.; Solich, P.; Spacil, Z.; Novakova, L.; Solich, P. Analysis of phenolic compounds by high performance liquid chromatography and ultra performance liquid chromatography. *Talanta* **2008**, *76*, 189–199. [CrossRef] [PubMed]
82. Natividade, M.M.P.; Corrêa, L.C.; Souza, S.V.C.; Pereira, G.E.; Lima, L.C.O. Simultaneous analysis of 25 phenolic compounds in grape juice for HPLC: Method validation and characterization of São Francisco Valley samples. *Microchem. J.* **2013**, *110*, 665–674. [CrossRef]
83. Xia, E.-Q.; Deng, G.-F.; Guo, Y.-J.; Li, H.-B. Biological Activities of Polyphenols from Grapes. *Int. J. Mol. Sci.* **2010**, *11*, 622–646. [CrossRef] [PubMed]
84. Soleas, G.J.; Diamandis, E.P.; Goldberg, D.M. Resveratrol: A molecule whose time has come? And gone? *Clin. Biochem.* **1997**, *30*, 91–113. [CrossRef]
85. Carbonaro, M.; Mattera, M.; Nicoli, S.; Bergamo, P.; Capelloni, M. Modulation of antioxidant compounds in organic *vs.* conventional fruit (peach, *Prunus persica* L.; and pear, *Pyrus commumis* L.). *J. Agric. Food Chem.* **2002**, *50*, 5458–5462. [CrossRef] [PubMed]
86. Grinder-Pederson, L.; Rasmussen, S.E.; Bugel, S.; Jørgensen, L.V.; Dragsted, L.O.; Gundersen, V.; Sandström, B. Effect of diets based on foods from conventional versus organic production on intake and

excretion o flavonoids and markers of antioxidative defense in humans. *J. Agric. Food Chem.* **2003**, *51*, 5671–5676. [CrossRef] [PubMed]

87. Leblanc, M.R.; Johnson, C.E.; Wilson, P.W. Influence of pressing method on juice stilbene content in Muscadine and Bunch grapes. *J. Food Sci.* **2008**, *73*, 58–62. [CrossRef] [PubMed]

88. Treptow, T.C.; Franco, F.W.; Mascarin, L.G.; Hecktheuer, L.H.R.; Sautter, C.K. Physicochemical Composition and Sensory Analysis of Whole Juice Extracted from Grapes Irradiated with Ultraviolet C. *Rev. Bras. Frutic.* **2017**, *39*, e579. [CrossRef]

89. Mattivi, F.; Guzzon, R.; Vrhovsek, U.; Stefanini, M.; Velasco, R. Metabolite profiling of grape: Flavonols and anthocyanins. *J. Agric. Food Chem.* **2006**, *54*, 7692–7702. [CrossRef] [PubMed]

90. Burin, V.M.; Falcão, L.D.; Gonzaga, L.V.; Fett, R.; Rosier, J.P.; Bordignon-Luiz, M.-T. Colour, phenolic content and antioxidant activity of grape juice. *Food Sci. Technol.* **2010**, *30*, 1027–1032. [CrossRef]

91. Gurak, P.D.; Cabral, L.M.C.; Rocha-Leão, M.H.M.; Matta, V.M.; Freitas, S.P. Quality evaluation of grape juice concentrated by reverse osmosis. *J. Food Eng.* **2010**, *96*, 421–426. [CrossRef]

92. Downey, M.O.; Krstic, M.K.; Dokoozlian, N.K. Cultural practice and environment impacts on the flavonoid composition of grapes and wine—A review of recent research. *Am. J. Enol. Vitic.* **2006**, *57*, 257–268.

93. Bautista-Ortín, A.B.; Fernández-Fernández, J.I.; López-Roca, J.M.; Gómez-Plaza, M. The effects of enological practices in anthocyanins, phenolic compounds and wine colour and their dependence on grape characteristics. *J. Food Compos. Anal.* **2007**, 546–552. [CrossRef]

94. Meullenet, J.-F.; Lovely, C.; Threlfall, R.; Morris, J.R.; Striegler, R.K. An ideal point density plot method for determining an optimal sensory profile for Muscadine grape juice. *Food Qual. Preference* **2008**, *19*, 210–219. [CrossRef]

95. Huckleberry, J.M.; Morris, J.R.; James, C.; Marx, D.; Rathburn, I.M. Evaluation of Wine Grapes for Suitability in Juice Production. *J. Food Qual.* **1990**, *13*, 71–84. [CrossRef]

96. Girard, B.; Mazza, G. Produtos Funcionales derivados de lãs uvas y de los cítricos. Cap 5. In *Alimentos Funcionales: Aspectos Bioquímicos e de Procesado*; Mazza, G., Acribia, S.A., Eds.; Acribia: Zaragoza, Spain, 1998; pp. 141–182.

97. Öncül, N.; Karabiyikli, S. Factors affecting the quality attributes of unripe grape functional food products. *J. Food Biochem.* **2015**, *39*, 689–695. [CrossRef]

98. Simone, G.V.; Montevecchi, G.; Masino, F.; Matrella, V.; Imazio, S.A.; Antonelli, A.; Bignami, C. Ampelographic and chemical characterization of Reggio Emilia and Modena (northern Italy) grapes for two traditional seasonings: 'Saba' and 'agresto'. *J. Sci. Food Agric.* **2013**, *93*, 3502–3511. [CrossRef] [PubMed]

99. Alipour, M.; Davoudi, P.; Davoudi, Z. Effects of unripe grape juice (verjuice) on plasma lipid profile, blood pressure, malondialdehyde and total antioxidant capacity in normal, hyperlipidemic and hyperlipidemic with hypertensive human volunteers. *J. Med. Plants Res.* **2012**, *6*, 5677–5683. [CrossRef]

100. Setorki, M.; Asgary, S.; Eidi, A.; Rohani, A.H. Effects of acute verjuice consumption with a high-cholesterol diet on some biochemical risk factors of atherosclerosis in rabbits. *Med. Sci. Monit.* **2010**, *16*, 124–130.

101. Öncül, N.; Karabiyikli, S. Survival of foodborne pathogens in unripe grape products. *LWT Food Sci. Technol.* **2016**, *74*, 168–175. [CrossRef]

102. Karabiyikli, Ş.; Öncül, N. Inhibitory effect of unripe grape products on foodborne pathogens. *J. Food Process. Preserv.* **2016**, *40*, 1459–1465. [CrossRef]

103. Soares, S.; Vitorino, R.; Osório, H.; Fernandes, A.; Venâncio, A.; Mateus, N.; Freitas, V. Reactivity of human salivary proteins families toward food polyphenols. *J. Agric. Food Chem.* **2011**, *59*, 5535–5547. [CrossRef] [PubMed]

104. Matos, A.D.; Curioni, A.; Bakalinsky, A.T.; Marangon, M.; Pasini, G.; Vincenzi, S. Chemical and sensory analysis of verjuice: An acidic food ingredient obtained from unripe grape berries. *Innov. Food Sci. Emerg. Technol.* **2017**, *44*, 9–14. [CrossRef]

105. Taste of Beirut. Sour Grapes to Make Verjuice or Husrum. Available online: http://www.tasteofbeirut.com/sour-grapes-to-make-verjuice-or-husrum/ (accessed on 12 December 2017).

beverages

MDPI

Article

Electrochemistry of White Wine Polyphenols Using PEDOT Modified Electrodes

Qiang Zhang [1], Alexander Türke [2] and Paul Kilmartin [1,*]

[1]   School of Chemical Sciences, The University of Auckland, Private Bag 92019, Auckland, New Zealand; qzha218@aucklanduni.ac.nz
[2]   ILK Dresden gGmbH, Bertolt-Brecht-Allee 20, 01309 Dresden, Germany; alexander.tuerke@ilkdresden.de

Academic Editor: António Manuel Jordão
Received: 1 June 2017; Accepted: 21 June 2017; Published: 28 June 2017

Abstract: The conducting polymer PEDOT (poly-3,4-ethylenedioxythiophene) has been polymerized onto 3 mm and 10 μm electrodes from a propylene carbonate solution. The electrodes have then been tested in a Chardonnay wine, including dilutions in a model wine solution, with comparisons made to scans with a glassy carbon electrode. A well-defined oxidation peak was obtained for the white wine at PEDOT in the 400 to 450 mV (Ag/AgCl) range, where peaks were also obtained for the representative phenolics caffeic acid and catechin. The voltammetry at PEDOT was typical of a surface-confined process. Significant preconcentration, leading to an increased current response, was noted over a period of 20 min of holding time. Extensive PEDOT growth was observed in the microelectrode case, leading to current densities for the oxidation of caffeic acid over 1000 times greater than those observed at the macroelectrode, matching the high surface area and fractal-type growth observed in SEM images.

Keywords: polyphenols; wine; cyclic voltammetry; PEDOT

---

1. Introduction

Electrochemical methods to characterize the phenolic content of wines have been applied widely over the past 15 years, mainly based upon cyclic voltammetric analyses at carbon-based electrodes [1–7]. Many polyphenols contain *ortho*-diphenol (catechol) or triphenol (galloyl) groups, which are important in beverage oxidation processes, given that they are they are recognized as the main initial substrate of wine oxidation [7], and they are oxidized at a relatively low electrode potential. Dilution of wines is generally required to improve peak resolution and to obtain a linear relationship between dilution factor and peak current. The removal of free sulfites, through addition of acetaldehyde, has been used to obtain a more reliable measure of the phenolic compounds present [7]. Under these conditions, the anodic current response can be related to levels of monomeric phenolics determined by HPLC, at least with white wines, where the phenolic composition is dominated by hydroxycinnamic acids and small flavonoids [1,7]. The advantages of the use of differential pulse voltammetry have also been reported, including improved peak resolution and a good correlation with spectrophotometric measures of red wine phenolics [8–10].

A further approach that has applied to the voltammetric analysis of wines has been to use a modified electrode. Poly-3,4-ethylenedioxythiophene (PEDOT) has been of particular interest [11–13], with improved sensitivity and antifouling properties. Separate current signals due to ascorbic acid and wine polyphenols have been obtained, which can merge at glassy carbon electrodes due to inhibitory effects of adsorbed polyphenols on ascorbic acid oxidation [14,15]. In further work with PEDOT electrodes, an improved response for tea catechins was obtained compared to glassy carbon electrodes [16], and the most responsive PEDOT films were those prepared originally in a propylene carbonate electrolyte [17].

At the same time, adsorption effects can be of increasing importance when high surface area modified electrodes are employed. In this report, the application of PEDOT electrodes to the characterization of polyphenols in white wines is presented, including the influence of holding-time in the sample solution prior to taking the voltammetric scan. High surface area PEDOT microelectrodes are also examined, with unique features being obtained.

2. Materials and Methods

2.1. Materials

Caffeic acid, (+)-catechin hydrate, 3,4-ethylenedioxythiophene (EDOT), propylene carbonate, and lithium perchlorate were purchased from Sigma-Aldrich. A 2008 Chardonnay white wine was selected for testing, a wine made within the University of Auckland Wine Science program under the Ingenio label, using grapes sourced from Waiheke Island in the Auckland region. The wine had been bottled under screwcap closure and was opened immediately prior to analysis in mid-2014.

2.2. Preparation of PEDOT Sensor

Electropolymerization of the PEDOT electrodes was undertaken on a 3 mm diameter glassy carbon electrode (BAS MF-2012; 0.0707 cm^2) in a solution of 0.1 M 3,4-ethylenedioxythiophene in propylene carbonate with 0.1 M LiClO$_4$, after the removal of oxygen by purging with nitrogen [13,16]. The glassy carbon electrodes were first polished using 0.05 μm alumina powder (BAS CF-1050) on a cloth pad. Four preparative potential sweeps were run between -300 and 1200 mV (Ag/AgCl, BAS MF-2052, $+207$ mV versus SHE) at 100 mV s^{-1} with a platinum wire as counter electrode in a Bioanalytical Systems Electrochemical Analyzer (BAS100A). The PEDOT electrodes were then cycled 10 times at 100 mV s^{-1} from -200 and 800 mV (Ag/AgCl), in a model wine solution, consisting of 0.033 M L-tartaric acid in a 12% (v/v) ethanol-water mixture with the pH adjusted to 3.6 ± 0.2 with 1 M NaOH.

A further PEDOT preparation was undertaken on a 10 μm gold microelectrode (BAS MF-2006; 7.85×10^{-7} cm^2). Between two and eight preparative sweeps were undertaken to 1200 mV (Ag/AgCl), with a sharp rise in current beyond 1100 mV on the first scan, and from 900 to 1000 mV on subsequent scans. SEM images of the PEDOT microelectrodes were taken after zero, two, and eight preparative cycles, using a Philips XL30S FEG Scanning Electron Microscope, equipped with a SiLi EDS detector [17].

2.3. Cyclic Voltammetry Measurements

The PEDOT electrodes were then transferred to the solution containing diluted wine or phenolic standard, which had been previously deaerated by purging with nitrogen, and the first voltammetric scan was recorded either immediately, or after a certain holding period. Cyclic voltammetry was performed between -200 mV and a final potential as high as 1200 mV, at a scan rate of 100 mV s^{-1}. Where repeat runs were undertaken ($n = 3$) and values such as the peak currents obtained, the averages with standard deviations are presented. A voltammogram was also taken in the model wine solution on its own for background subtraction purposes. For comparison purposes, additional scans for the diluted white wine were undertaken with a clean glassy carbon electrode in the absence of a PEDOT film. In each case, both the raw current values and the current density with respect to the geometric surface area of the underlying inert electrode are presented.

Solutions of the pure phenolic compounds, caffeic acid and catechin were also prepared in the model wine solution at a range of concentrations from 0.01 to 0.2 mM, and tested at the PEDOT electrodes. For each concentration value, a freshly prepared PEDOT film was used.

3. Results and Discussion

3.1. Cyclic Voltammetry of White Wines

Voltammograms were recorded both at PEDOT and glassy carbon electrodes for the Chardonnay wine, and for successive dilutions in the model wine solution (Figure 1). The scans were taken to 1200 mV (Ag/AgCl) to cover the full range of oxidizable phenolics expected to be present in the wine, and in both cases the background response due to the PEDOT or glassy carbon electrode in the model wine solution has been subtracted away. The rising current for potentials greater than 350 mV is mainly due to the presence of catechol-containing phenolics such as caffeic acid derivatives and catechin, with a further contribution from free SO_2 also possible [7]. However, with an undiluted wine at a glassy carbon electrode, the current lacks definition into well-resolved peaks, and after a scan to 1200 mV, the presence of oxidized phenolic products on the electrode also limits the appearance of a quinone reduction peak expected on the reverse scan [1]. By contrast, at the PEDOT electrode, a clear peak was observed at 460 mV in the pure wine, and at close to 410 mV for the 10-fold diluted wine, while after cycling to 1200 mV, a reduction peak at 320 to 350 mV was also seen. A further large current peak at 900–1000 mV can be associated with more isolated phenolic groups, such as coutaric acid, while a smaller peak or shoulder feature can be noted at around 700 mV. Furthermore, the intensity of the current was around five times greater at the PEDOT electrode compared to the bare glassy carbon electrode.

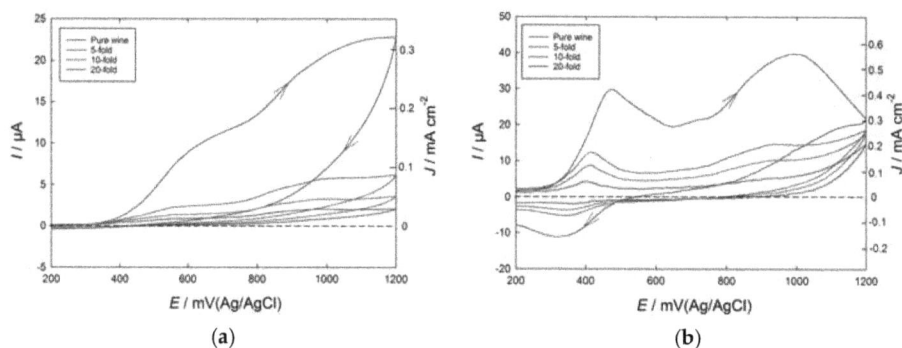

Figure 1. Cyclic voltammograms of a Chardonnay wine, both pure and diluted successively, in a pH 3.6 model wine solution: (**a**) scans taken at a bare glassy carbon electrode; (**b**) scans taken at a PEDOT electrode. All of the voltammograms were measured at 100 mV s^{-1} on 3 mm diameter electrodes, and are presented after subtracting away the voltammogram obtained for a blank run.

3.2. Cyclic Voltammetry of Phenolic Standards

Further voltammograms at PEDOT electrodes were run on the representative catechol-containing hydroxycinnamic acid, caffeic acid (Figure 2a), and the flavonoid, catechin (Figure 2b). The standards were tested from 0.01 to 0.20 mM, and a well-defined set of redox peaks were obtained in each case in the 300 to 500 mV potential range, due to the oxidation of the catechol group to a quinone-form, and its reduction on the reverse scan. The mid-point potentials were, on average, 422 ± 2 mV for caffeic acid and 386 ± 1 mV for catechin, close to those reported previously for these phenolic compounds in a pH 3.6 model wine solution [1]. In each case, the peak separation decreased as the concentration of the phenolics was lowered from 0.20 mM to 0.01 mM, moving from 120 to 52 mV for the caffeic acid voltammograms, and from 88 to 37 mV for catechin, while the mid-point potentials were quite steady. The height of the current peak for 0.050 mM caffeic acid or catechin, at 10.2 and 9.2 µA, respectively, was comparable to the value of 8.8 µA recorded for the 10-fold diluted Chardonnay wine, which by this measure would contain just over 0.5 mM of such phenolics in the undiluted wine (c. 100 mg/L caffeic acid equivalents). This value is typical of those obtained previously for voltammetric quantification of the phenolics in white wines that can be oxidized at a potential less than 500 mV [18].

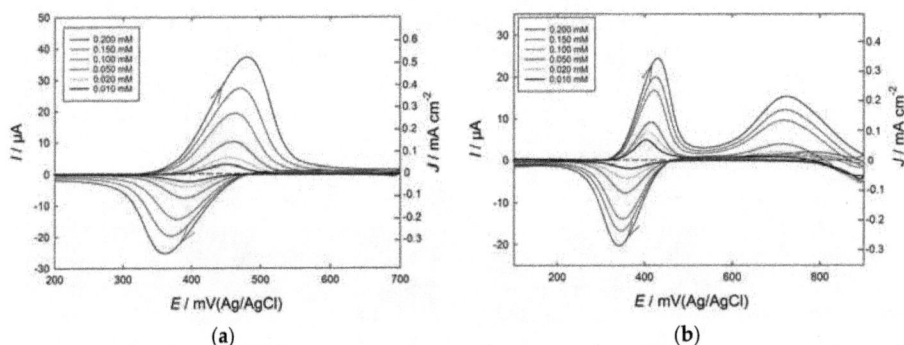

Figure 2. Cyclic voltammograms at a PEDOT electrode of phenolic standards in the concentration range from 0.01 to 0.20 mM, dissolved in the pH 3.6 model wine solution: (**a**) caffeic acid taken to 700 mV; (**b**) catechin taken to 900 mV. All of the voltammograms were measured at 100 mV s^{-1} and are presented after subtracting away the voltammogram obtained for a blank run.

At the same time, the oxidation of caffeic acid was largely completed by 600 mV, at which point the current returned to the background response due to the PEDOT conducting polymer itself (Figure 2a), and without the ongoing current expected for a process under diffusion-control typical of caffeic acid at a glassy carbon electrode [14]. Instead, the oxidation and subsequent reduction peaks point to the electrochemistry of an absorbed species. At the same time, the peak separation between anodic and cathodic peaks was non-zero, as would be expected for a fully-Nernstian response, pointing to some kinetic limitations and partial irreversibility in the electrochemistry. In the case of catechin, the additional current from 600 mV, peaking at around 720 mV, is expected due to the oxidation of the extra meta-diphenol groups on the flavonoid A-ring. The anodic current seen in the 600 to 800 mV range at the PEDOT electrode in the Chardonnay wine (Figure 1b) could also derive from this source. For catechin concentrations less than 0.10 mM, the current by 900 mV was in fact lower than the background current due to PEDOT run in the absence of catechin (Figure 2b), giving rise to negative currents at potentials between 800 and 900 mV, where the effects of adsorbed catechin oxidation products may be affecting PEDOT electrochemistry.

3.3. Adsorption Effects at PEDOT Electrodes

To investigate the aspects of phenolic adsorption on PEDOT and glassy carbon electrodes, the dependence of the first anodic peak current upon scan rate was examined for the situation in which the voltammetry was run immediately after inserting the electrode into the test solution. For 0.02 mM caffeic acid at the glassy carbon electrode, a log/log plot indicated that the behavior was close to that expected for a process under diffusion control, with a slope of 0.593 close the value of 0.5, obtained when the peak current is proportional to the square root of the sweep rate (Figure 3). In the case of the PEDOT electrode, a slope of 0.829 for 0.02 mM caffeic acid and 0.857 for the 10-fold diluted Chardonnay wine was closer to the value of 1.0 expected for a surface confined process with peak current directly proportional to scan rate. Such behavior is typical of an absorbed layer of redox active species, while some smaller diffusional component may serve to bring the value down from 1.0.

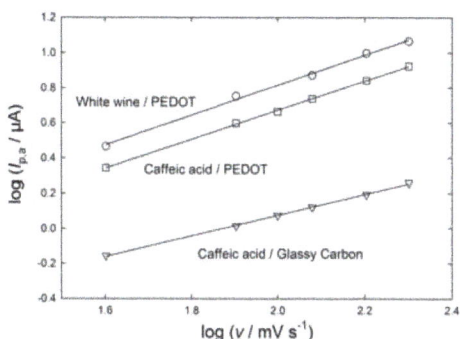

Figure 3. Logarithm of the first anodic peak current ($I_{p,a}$) versus logarithm of sweep rate (v) for the Chardonnay wine diluted 10-fold, and for 0.02 mM caffeic acid, both prepared in the pH 3.6 model wine solution, taken at bare glassy carbon and PEDOT electrodes immediately after the electrodes were inserted into the test solution.

The role of adsorption at PEDOT electrodes was further tested by introducing a holding time from 3 to 20 min for the electrode immersed in the 10-fold diluted Chardonnay wine (Figure 4a). A considerable increase in current was seen as the holding time progressed, with a value close to the maximum being reached after 10 min of holding. The current dependence upon holding time for three wine-dilution levels is shown in Figure 4b, and the effects of caffeic acid pre-concentration can be seen in each case, which resemble the isotherms observed with the adsorption of species at surfaces. There are benefits to enhancing the signal by introducing a holding time, but in the present case this came at the expense of a loss of the linear relationship between wine dilution, and hence concentration of phenolic compounds, and peak current. For a scan run immediately after inserting the PEDOT electrodes into the diluted wines, the peak current nearly doubled, moving from 20-fold to 10-fold diluted wines ($\times 1.8$), and increased by a similar amount with the 5-fold diluted wines ($\times 1.9$). By contrast, with a 20 min holding time, the response to the 20-fold diluted wine was over 50% of that of the 5-fold diluted wine.

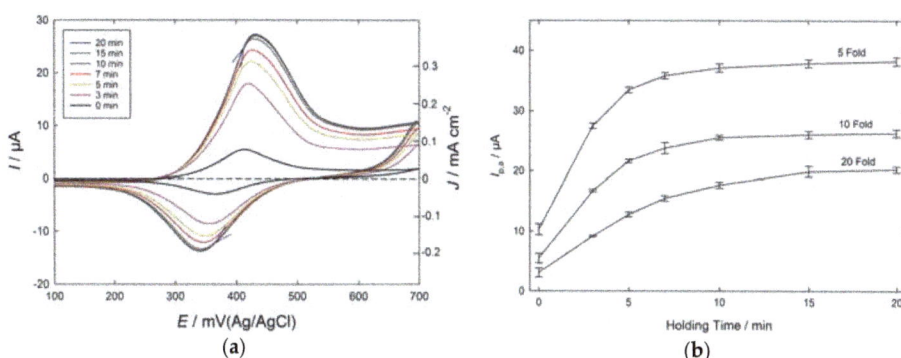

Figure 4. Voltammetric response of PEDOT electrodes taken at 100 mV s^{-1} of the Chardonnay wine diluted in a pH 3.6 model wine solution for different holding times: (**a**) voltammograms (background subtracted) obtained for 10-fold diluted wine; (**b**) first anodic peak current ($I_{p,a}$) for three different wine dilution levels at different holding times ($n = 3$).

The analysis of phenolics on PEDOT electrodes after an accumulation time has also been described for the case of epicatechin, as found in products such as biscuits containing cocoa [19]. After a period of accumulation, the oxidation of the phenol then occurs as an example of adsorptive stripping voltammetry, with complete oxidation of the adsorbed phenolic compound, and with a current much higher than that obtained at glassy carbon electrodes. In that study, the current peak for epicatechin reached a maximum after 25 min of accumulation time, and showed a linear response for epicatechin concentrations up to 1.5 ppm (2.0 µA for c. 0.005 mM) using differential pulse voltammetry, with linearity lost beyond 2.5 ppm [19]. A similar accumulation period was found to provide maximum oxidation current in the present study for white wine phenols.

3.4. Cyclic Voltammetry of Caffeic Acid at PEDOT Microelectrodes

PEDOT electrodes were also prepared on a 10 µm diameter gold microelectrode, and used to test solutions of caffeic acid. After two preparative cycles to 1200 mV (Ag/AgCl) for 0.1 M EDOT in propylene carbonate, some PEDOT growth nodules were evident in SEM images (Figure 5a). The PEDOT conducting polymer had also spread to a width of around 25 µm, somewhat larger than the underlying gold microelectrode, which had been previously found to be very close to the expected 10 µm in SEM images of the bare microelectrode. During the electropolymerization, the current for EDOT oxidation rose sharply for potentials greater than 1100 mV on the first preparative scan, but by the third cycle the rise in current started from 900 mV, pointing to an electrocatalytic effect of the newly formed PEDOT surface. The extremely high rate of growth was most evident in the following cycles, and for continued cycling even to 16 prepared scans, a steady increase in current due to internal PEDOT redox processes was observed. A SEM image of a PEDOT electrode formed by eight preparative scans to 1200 mV is shown in Figure 5b, and the extreme nature of the fractal-type growth under these conditions is evident. At this point, the PEDOT material extended some 50 µm out from the electrode surface, anchored by an area about 25 µm wide.

(a)   (b)

Figure 5. Scanning Electron Microscopy images of PEDOT growth on a 10 µm gold microelectrode, after (a) two, and (b) eight electropolymerization cycles taken to 1200 mV (Ag/AgCl).

When solutions of caffeic acid were tested at the PEDOT microelectrodes (Figure 6), a similar-shaped anodic peak was obtained at 400 to 500 mV, as seen at the 3 mm diameter PEDOT macroelectrode (Figure 2a). Should the response have been under diffusion control at the microelectrode, a plateau current at potentials greater than 400 mV would be expected. However, the current density was now greatly enhanced, and provides a further example of adsorptive stripping voltammetry. For the case of the macroelectrode with 0.02 mM caffeic acid, the peak current of 4 µA at the 0.071 cm^2 electrode was thus 0.057 mA cm^{-2}. By contrast, a current of 230 nA for 0.022 mM caffeic acid (4 ppm), at the 7.9×10^{-7} cm^2 electrode was 290 mA cm^{-2}, over 1000 times larger than the macroelectrode case. These 'apparent' current densities are reported in terms of the geometric surface are of the underlying glassy carbon or gold electrodes, but it is clear that the effective electrode area of PEDOT exposed to the solution in the microelectrode case is much larger, given the spherical shapes of the PEDOT deposits seen in SEM images. With the 3 mm diameter electrode, if further preparative scans with the PEDOT monomer were applied out to eight scans in total, the current response for caffeic acid did not increase significantly beyond that obtained with a four scan preparation, which was considered an optimum for the macroelectrode.

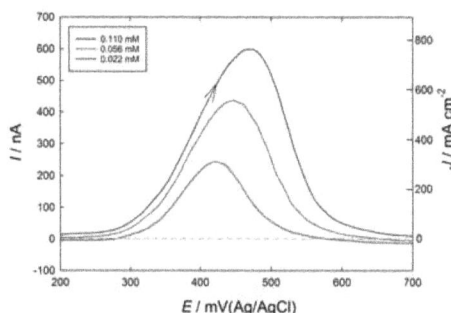

Figure 6. Voltammograms at PEDOT grown on a 10 μm gold microelectrode by eight preparative cycles to 1200 mV (Ag/AgCl), for caffeic acid in the concentration range from 0.022 to 0.110 mM, dissolved in the pH 3.6 model wine solution. The voltammograms were measured at 100 mV s^{-1} and are presented after subtracting away the voltammogram obtained for a blank run, with only the forward anodic scan shown.

Despite the very large current densities involved at the PEDOT microelectrode, the current value had decreased back close to the background PEDOT response as the potential reached 600 mV, showing that high conductivity of the PEDOT conducting polymer had allowed the caffeic acid adsorbed throughout the PEDOT structure to be oxidized over the 3 s time period of the voltammetric scan, passing from 300 to 600 mV. Comparisons can be drawn with the high surface area response obtained with platinized platinum, compared to bright polished platinum, where roughness factors of the orders of several hundreds have been obtained [20].

4. Conclusions

PEDOT electrodes show considerable promise for applications in electroanalytical chemistry, based upon their responsiveness to oxidizable substrates, such as wine polyphenols, and the separation of signals obtained when other reducing agents, such as ascorbic acid, are present. Considerable pre-concentration of phenolic species occurs at PEDOT electrodes, which can be followed by a stripping voltammogram for the quantification of phenolics present in beverages such as wines and teas. The very high surface area afforded by certain PEDOT preparations at microelectrodes raises the possibility of developing highly sensitive microelectrode systems. Further research is needed to optimize experimental conditions, including holding time, along with further correlations with phenolic composition by independent measures such as HPLC.

Author Contributions: Alexander Türke and Paul Kilmartin conceived and designed the experiments; Qiang Zhang and Alexander Türke performed the experiments and analyzed the data; Qiang Zhang and Paul Kilmartin wrote the paper.

Conflicts of Interest: The authors declare no conflict of interest.

References

1. Kilmartin, P.A.; Zou, H.; Waterhouse, A.L. A cyclic voltammetry method suitable for characterizing antioxidant properties of wine and wine phenolics. *J. Agric. Food Chem.* **2001**, *49*, 1957–1965. [CrossRef] [PubMed]
2. Kilmartin, P.A.; Zou, H.; Waterhouse, A.L. Correlation of wine phenolic composition versus cyclic voltammetry response. *Am. J. Enol. Vitic.* **2002**, *53*, 294–302.
3. Piljac, J.; Martinez, S.; Stipčević, T.; Petrović, Z.; Metikoš-Huković, M. Cyclic voltammetry investigation of the phenolic content of Croatian wines. *Am. J. Enol. Vitic.* **2004**, *55*, 417–422.
4. Rodrigues, A.; Ferreira, A.C.S.; de Pinho, P.G.; Bento, F.; Geraldo, D. Resistance to oxidation of white wines assessed by voltammetric means. *J. Agric. Food Chem.* **2007**, *55*, 10557–10562. [CrossRef] [PubMed]
5. Kzenzhek, O.; Petrova, S.; Kolodyazhny, M. Redox Spectra of wines. *Electroanalysis* **2007**, *19*, 389–392. [CrossRef]
6. Petrovic, S.C. Correlation of perceived wine astringency to cyclic voltammetric response. *Am. J. Enol. Vitic.* **2009**, *60*, 373–378.
7. Makhotkina, O.; Kilmartin, P.A. The use of cyclic voltammetry for wine analysis: Determination of polyphenols and free sulfur dioxide. *Anal. Chim. Acta.* **2010**, *668*, 155–165. [CrossRef] [PubMed]
8. Aguirre, M.J.; Chen, Y.Y.; Isaacs, M.; Matsuhiro, B.; Mendoza, L.; Torres, S. Electrochemical behavior and antioxidant capacity of anthocyanins from Chilean red wine, grape and raspberry. *Food Chem.* **2010**, *121*, 44–48. [CrossRef]
9. Seruga, M.; Novak, I.; Jakobek, L. Determination of polyphenols content and antioxidant activity of some red wines by differential pulse voltammetry, HPLC and spectrophotometric methods. *Food Chem.* **2011**, *124*, 1208–1211. [CrossRef]
10. Rebelo, M.J.; Rego, R.; Ferreira, M.; Oliveira, M.C. Comparative study of the antioxidant capacity and polyphenol content of Douro wines by chemical and electrochemical methods. *Food Chem.* **2013**, *141*, 566–573. [CrossRef] [PubMed]
11. Pigani, L.; Foca, G.; Ionescu, K.; Martina, V.; Ulrici, A.; Terzi, F.; Vignali, M.; Zanardi, C.; Seeber, R. Amperometric sensors based on poly (3,4-ethylenedioxythiophene)-modified electrodes: Discrimination of white wines. *Anal. Chim. Acta.* **2008**, *614*, 213–222. [CrossRef] [PubMed]
12. Pigani, L.; Foca, G.; Ulrici, A.; Ionescu, K.; Martina, V.; Terzi, F.; Vignali, M.; Zanardi, C.; Seeber, R. Classification of red wines by chemometric analysis of voltammetric signals from PEDOT-modified electrodes. *Anal. Chim. Acta.* **2009**, *643*, 67–73. [CrossRef] [PubMed]
13. Türke, A.; Fischer, W.-J.; Beaumont, N.; Kilmartin, P. Electrochemistry of sulfur dioxide, polyphenols and ascorbic acid at poly (3, 4-ethylenedioxythiophene) modified electrodes. *Electrochim. Acta.* **2012**, *60*, 184–192. [CrossRef]
14. Makhotkina, O.; Kilmartin, P.A. Uncovering the influence of antioxidants on polyphenols oxidation in wines using an electrochemical method: Cyclic voltammetry. *J. Electroanal. Chem.* **2009**, *633*, 165–174. [CrossRef]
15. Bianchini, C.; Curulli, A.; Pasquali, M.; Zane, D. Determination of caffeic acid in wine using PEDOT film modified electrode. *Food Chem.* **2014**, *156*, 81–86. [CrossRef] [PubMed]
16. Karaosmanoglu, H.; Travas-Sejdic, J.; Kilmartin, P.A. Designing PEDOT-based sensors for antioxidant analysis. *Int. J. Nanotech.* **2014**, *11*, 445–450. [CrossRef]
17. Karaosmanoglu, H.; Travas-Sejdic, J.; Kilmartin, P.A. Comparison of organic and aqueous polymerized PEDOT sensors. *Mol. Cryst. Liq. Cryst.* **2014**, *604*, 233–239. [CrossRef]
18. Beer, D.D.; Harbertson, J.F.; Kilmartin, P.A.; Roginsky, V.; Barsukova, T.; Adams, D.O.; Waterhouse, A.L. Phenolics: A comparison of diverse analytical methods. *Am. J. Enol. Vitic.* **2004**, *55*, 389–400.
19. Pigani, L.; Seeber, R.; Bedini, A.; Dalcanale, E.; Suman, M. Adsorptive-stripping voltammetry at PEDOT-modified electrodes. Determination of epicatechin. *Food Anal. Methods* **2014**, *7*, 754–760. [CrossRef]
20. Feltham, A.M.; Spiro, M. Platinized platinum electrodes. *Chem. Rev.* **1971**, *71*, 177–193. [CrossRef]

beverages

MDPI

Article

Wine Phenolic Compounds: Antimicrobial Properties against Yeasts, Lactic Acid and Acetic Acid Bacteria

Andrea Sabel, Simone Bredefeld, Martina Schlander and Harald Claus *

Microbiology & Wine Research, Institute of Molecular Physiology, Johannes Gutenberg-University, Mainz 55099, Germany; andrea-sabel@gmx.de (A.S.); simonebredefeld@googlemail.com (S.B.); schlande@uni-mainz.de (M.S.)

Academic Editor: António Manuel Jordão
Received: 24 April 2017; Accepted: 23 June 2017; Published: 29 June 2017

Abstract: Microorganisms play an important role in the conversion of grape juice into wine. Yeasts belonging the *genus Saccharomyces* are mainly responsible for the production of ethanol, but members of other genera are known as producers of off-flavors, e.g., volatile phenols. Lactic acid and acetic acid bacteria also occur regularly in must and wine. They are mostly undesirable due to their capacity to produce wine-spoiling compounds (acetic acid, biogenic amines, *N*-heterocycles, diacetyl, etc.). In conventional winemaking, additions of sulfite or lysozyme are used to inhibit growth of spoilage microorganisms. However, there is increasing concern about the health risks connected with these enological additives and high interest in finding alternatives. Phenols are naturally occurring compounds in grapes and wine and are well known for their antimicrobial and health-promoting activities. In this study, we tested a selection of phenolic compounds for their effect on growth and viability of wine-associated yeasts and bacteria. Our investigations confirmed the antimicrobial activities of ferulic acid and resveratrol described in previous studies. In addition, we found syringaldehyde highly efficient against wine-spoiling bacteria at concentrations of 250–1000 μg/mL. The promising bioactive activities of this aromatic aldehyde and its potential for winemaking deserves further research.

Keywords: wine; microorganisms; phenols; sulfite; growth inhibition; syringaldehyde; resveratrol; laccase

---

1. Introduction

Yeasts, lactic acid bacteria and acetic acid bacteria can play a role in winemaking and have an influence on wine quality [1]. Bacterial growth in musts and wines is conventionally controlled by the addition of sulfur dioxide. Unfortunately, the presence of sulfites in alcoholic beverages, particularly in wines, can cause pseudo-allergic responses with symptoms ranging from gastrointestinal problems to anaphylactic shock [2,3]. Other antimicrobials such as sorbic acid and dimethyl dicarbonate are active against yeasts, but have less activity against bacteria [2,3].

Phenolic compounds affect bacterial growth and metabolism [2]. Enological extracts have been found to inhibit clinically important microorganisms like *Staphylococcus aureus*, *Escherichia coli*, *Bacillus cereus*, *Campylobacter jejuni*, *Salmonella infantis* and *Candida albicans* [4–7]. Phenols kill microorganisms or inhibit the growth of bacteria, fungi or protozoa [8]. Several mechanisms of action in the growth inhibition of bacteria are involved, such as permeabilization and destabilization of the plasma membrane, or inhibition of extracellular microbial enzymes [2,9].

García-Ruiz et al. [10] investigated the inhibitory potential of 18 phenolic compounds (including hydroxybenzoic acids, hydroxycinnamic acids, phenolic alcohols, stilbenes, flavan-3-ols and flavonols) on wine-related lactic acid bacteria of the species *Oenococcus oeni*, *Lactobacillus hilgardii* and *Pediococcus pentosaceus*. In general, flavonols and stilbenes showed the greatest inhibitory effects on the growth of the test strains. Hydroxycinnamic acids and hydroxybenzoic acids and their esters exhibited medium inhibitory effect, and phenolic alcohols as well as flavanol-3-ols showed the lowest effect on the growth of the strains studied. In comparison to the antimicrobial additives used in winemaking, the inhibitory impact of most phenolic compounds was stronger than those of the commonly used potassium metabisulfite and lysozyme.

In a subsequent study, García-Ruiz et al. [11] determined whether phenolic extracts with antimicrobial activity may be considered as an alternative to the use of sulfur dioxide (SO_2) for controlling malolactic fermentation in winemaking. Growth inhibition of six enological strains of *Lactobacillus hilgardii, Lactobacillus casei, Lactobacillus plantarum, Pediococcus pentosaceus* and *Oenococcus oeni* by phenolic extracts from different origins (spices, flowers, leaves, fruits, legumes, seeds, skins, agricultural by-products and others) was evaluated. A total of 24 extracts were found to significantly inhibit the growth of at least two of the strains studied. Some of these extracts were also active against acetic acid bacteria.

In a similar approach, Pastorkova et al. [12] investigated the antimicrobial potential of 15 grape phenolic compounds of various chemical classes (phenolic acids, stilbenes and flavonoids) against yeasts and acetic acid bacteria. Pterostilbene, resveratrol and luteolin were among six active compounds that possessed the strongest inhibitory effects against all microorganisms tested. In the case of phenolic acids, myricetin, p-coumaric and ferulic acids exhibited selective antimicrobial activity depending upon the species of yeasts and bacteria tested. In comparison with potassium metabisulfite, all microorganisms tested were more susceptible to the phenols.

Recent studies suggested the possible application of plant phenolic extracts to reduce the use of sulfites in winemaking [13]. The use of polyphenols as a SO_2 substitute thus presupposes that wine yeasts are not influenced by the phenols in their fermentation activity, and only undesired organisms are inhibited. Hitherto, many studies have been concerned with the antimicrobial action of phenols against lactic acid bacteria but less with acetic acid bacteria and yeasts. The challenge is to find differences in the inhibitory effect of phenols within the various desirable and undesirable species.

2. Materials and Methods

2.1. Microorganisms and Culture Conditions

The yeasts *Wickerhamomyces anomalus* AS1, *Saccharomyces bayanus* HL 77, *Saccharomyces cerevisiae* × *Saccharomyces kudriavzevii* × *Saccharomyces* HL 78 and *Saccharomyces cerevisiae* 16.1 have been isolated from local wineries in Germany and characterized as described previously [14–16]. *Debaryomyces hansenii* 525 has been isolated from garden soil [17]. Strains of lactic acid bacteria (*Lactobacillus hilgardii* 20166, *Lactobacillus plantarum* 20174[T], *Pediococcus parvulus* 20332[T], *Oenococcus oeni* 20252) and acetic acid bacteria (*Gluconobacter cerinus* 9533, *Acetobacter acetii* 3508) were obtained from the German Collection of Microorganisms and Cell Cultures (DSMZ, Braunschweig).

Yeasts were cultivated in Sabouraud-glucose broth (DSMZ medium 1429) at pH 5.5 or 3.5 depending on the specific experimental conditions. Lactic acid bacteria (were cultivated in MRS medium (pH 5.5) containing tomato juice (DSMZ medium 11). *G. cerinus* 9533 was grown in YPM broth (DSMZ medium 360, pH 6.1) and *A. acetii* 3508 in YPS broth (with sorbitol instead of mannitol in YPM, pH 6.6). All cultures were incubated on a shaker at 30 °C.

2.2. Chemicals

All reagents were of analytical grade. Polydatin, caffeic acid, ferulic acid, p-coumaric acid, sinapic acid, 3,4-dihydroxybenzoic acid, gallic acid, ethylgallate, vanillic acid, syringaldehyde, and DMSO were purchased from Sigma-Aldrich (Taufkirchen, Germany). Resveratrol and microbiological culture media or constituents were supplied by Roth (Karlsruhe, Germany). Laccase from *Botrytis cinerea* strain P16–14 was prepared as described recently [18].

2.3. Influence of Ethanol and DMSO on Yeast Growth

The phenolic compounds were prepared in organic solvents. Their possible impact on yeast growth was tested as follows: Sabouraud-glucose broth media (pH 3.5) were prepared with 0, 1, 5, 10, 15 and 20% v/v ethanol and DMSO, respectively. The nutrient solutions were inoculated at initial cell numbers of 10^6/mL and incubated at 30 °C on a shaker (75 rpm) for four days. Growth of the cultures was recorded daily by determination of their optical densities (OD_{590}) for 4 days.

2.4. Influence of Phenols on Growth of Microorganisms

The effect of phenols on the growth of microbial liquid cultures was tested at different concentrations by monitoring the increase of cell densities for several days.

2.4.1. Method 1

The first experimental series was conducted using common glass culture tubes. Phenolic compounds were dissolved at concentrations of 5, 10 and 20 mg/mL in DMSO (99.4% v/v) and sterile-filtrated (0.2 µm). From each solution, 210 µL were added separately to 3.99 mL autoclaved Sabouraud-glucose broth (pH 3.5) to achieve final phenol concentrations of 250, 500 and 1000 µg/mL. The control media were prepared with 5% (v/v) DMSO. The inhibition effect of potassium metabisulfite on microbial cultures was compared at the same concentrations as used for the phenolic compounds.

The media were inoculated at initial yeast cell numbers of 10^6/mL and incubated at 30 °C on a shaker (75 rpm). Growth of the cultures was monitored by determination of their optical densities (OD_{590}) after 3 or 4 days with a Colorimeter model 45 (Fisher Scientific, Cambridge, UK). The optical densities of the DMSO controls were defined as 100% in relation to growth under the influence of phenolic compounds. The results are presented as the means of duplicate assays.

2.4.2. Method 2

The second experimental series was conducted in sterile 96-well plastic microplates (Roth, Karlsruhe, Germany). The phenolic compounds were dissolved at concentrations of 5, 10 and 20 mg/mL in DMSO (50% v/v) and sterile-filtrated (0.2 µm). Ten µL of each concentration and the DMSO control were added to 190 µL precultures of exponentially growing microorganisms (yeasts: 1×10^5 cells/ml; bacteria 1×10^6 cells/mL). The final concentrations of phenols were 250, 500 and 1000 µg/mL. The microplates were incubated in a wet chamber at 30 °C, acetic acid bacteria on a shaker. Growth was monitored by measuring the OD_{590} spectrophotometrically using a FLUOstar Omega microplate reader (BMG Labtech GmbH, Germany). The time of evaluation was set when the DMSO control reached the middle of the exponential growth phase. The corresponding optical densities were defined as 100% in relation to growth under the influence of phenolic compounds. The results are presented as the means of triplicate assays.

2.5. Influence of Laccase Oxidation on the Inhibition Effect of Phenolic Compounds

Abiotic and/or enzymatic oxidation may alter structure and reactivity of phenols; particularly laccase, a phenoloxidase produced by the wine-relevant fungus *Botrytis cinerea* plays an important role in this respect. For this reason, we tested the effect of enzymatic oxidations on the antimicrobial activities of selected phenols. The compounds were dissolved at concentrations of 5, 10 and 20 mg/mL in DMSO (50% v/v) and sterile-filtrated (0.2 µm). To the solutions 0.03 units/mL laccase of *Botrytis cinerea* P16–14 was added and incubation followed for 48 h at 30 °C. The controls contained no phenoloxidase. The next steps were performed as described above for method 2.

3. Results

3.1. Effect of Solvents

The phenolic compounds tested in this study are poorly soluble in water. For this reason, stock solutions had to be prepared in organic solvents such as dimethoxy sulfoxide (DMSO) or ethanol. As these solvents could have an impact on the inhibition experiments, we investigated the effect of different concentrations on the growth of the test microorganisms (Figure 1).

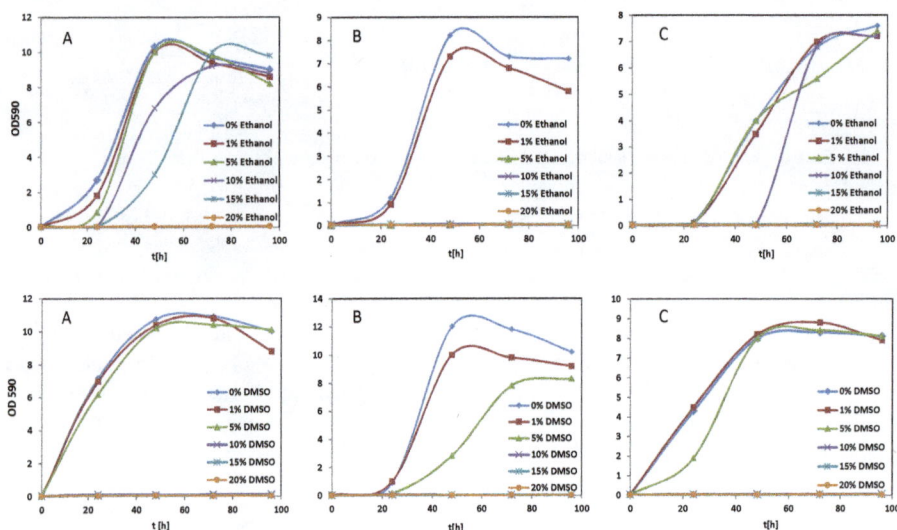

Figure 1. Influence of ethanol and DMSO on the growth of (**A**): *S. cerevisiae* 16.1; (**B**): *S. bayanus* HL 77; (**C**): *S. cerevisiae* × *S. kudriavzevii* × *S. bayanus* HL 78. Culture medium: Sabouraud-glucose broth (pH 3.5).

The Saccharomycetes exhibited quite different tolerances towards ethanol in the culture media. The most sensitive strain was *S. bayanus* HL 77, already 1% v/v alcohol inhibited growth. The most resistant *S. cerevisiae* 16.1 showed still—although a delayed—growth at 15% v/v, whereas the triple hybrid HL 78 was completely inhibited by this concentration (Figure 1, top row). A higher ethanol tolerance was expected for the wine yeast *S. bayanus* HL 77; however, it should be considered that in natural must fermentations yeasts have the chance for gradual adaptation to increasing ethanol concentrations, whereas in our experiments high amounts were already present from the start of cultivation. *W. anomalus* AS1 showed delayed growth at 10% ethanol in the medium and was completely inhibited by 15% and *Debaryomyces hansenii* 525 tolerated only 1% (not shown).

Regarding DMSO (Figure 1, bottom row), *S. cerevisiae* 16.1 showed no restricted growth in presence of 5% v/v, but complete failure at 10% v/v. This finding is in accordance with results of Sadowska-Bartosz et al. [19] reporting an inhibiting concentration of 5–10% v/v DMSO for this species. *S. bayanus* HL 77 showed delayed growth at 5% v/v, but reached still high cell densities after 4 days. Again, 10% v/v was not tolerated by this strain. This concentration was also the limit for the triple hybrid HL 78, which showed nearly normal growth behavior at 5% v/v. *W. anomalus* AS1 was not inhibited up to 5% v/v DMSO, whereas cell densities of *D. hansenii* 525 were drastically reduced at this concentration (not shown).

Although yeast growth, with the exception of strain 525, was generally not significantly reduced at 5% v/v DMSO, a second experimental series was performed with only half of this concentration.

At 2.5% v/v, also lactic acid and acetic acid bacteria reached the same cell densities as the controls without DMSO. Only *A. acetii* 3508[T] was more susceptible, the maximum OD_{595} was only 60% of the control (not shown).

3.2. Influence of Phenols on Yeast Growth

Different classes of phenolic compounds were tested for their antimicrobial activities (Table 1).

The first experimental series was performed at 5% v/v DMSO in the culture medium of pH 3.5 (Figure 2).

Growth of *S. cerevisiae* 16.1 was generally not negatively impaired by phenols. Only high concentrations of ferulic acid or potassium sulfite led to complete inhibition. In contrast, *S. bayanus* HL 77 and the triple hybrid were depressed by all phenols, even at low concentrations of *p*-coumaric acid and ferulic acid.

Table 1. Phenolic compounds included in this study and their natural concentrations in wine.

Chemical Class	Representative Structures	Examples	Mean Contents [μg/mL] [1] in	
			Red Wine	White Wine
Hydroxybenzoic acids		Gallic acid	35.9	2.2
		Vanillic acid	3.2	0.4
		4-Hydroxybenzoic acid	5.5	0.2
		Syringic acid	2.7	<0.01
		Ethylgallate	15.3	nd
Cinnamic acids		*p*-Coumaric acid	5.5	1.5
		Caffeic acid	18.8	2.4
		Ferulic acid	0.8	0.9
		Sinapic acid	0.7	0.6
Stilbenes		*cis*-Resveratrol	1.3	0.2
		Trans-Resveratrol	1.8	0.4
		trans-Resveratrol-3-O-glucoside	4.1	1.7
Hydroxybenz-aldehydes		Syringaldehyde	6.6	<0.01

[1] Source: http://phenol-explorer.eu; nd: no data.

The red line indicates maximum growth (optical density) of the controls which contained DMSO [5.0% v/v].

W. anomalus AS1 and *D. hansenii* 525 were sensitive to low concentrations of sulfite, ferulic acid and syringaldehyde. On the other hand, gallic acid stimulated growth of both non-Saccharomycetes.

The second experimental series was performed at 2.5% v/v DMSO in the culture medium of pH 5.5 (Figure 3).

Growth of *S. cerevisiae* 16.1 was completely inhibited by high concentrations of sulfite, resveratrol and ferulic acid and significantly impaired by syringaldehyde. The residual six phenols had no effect. The fact that potassium sulfite was not as effective as in the previous experiments can be explained in a diminished concentration of the effective molecular form (SO_2) at pH 5.5 compared to pH 3.5. *S. bayanus* HL 77 showed a very similar pattern, again resveratrol, ferulic acid and syringaldehyde were the most antimicrobial. Under the given experimental conditions, both non-Saccharomycetes turned out to be rather resistant to most phenols. However, a strong negative effect was exerted by resveratrol, followed by syringaldehyde and ferulic acid.

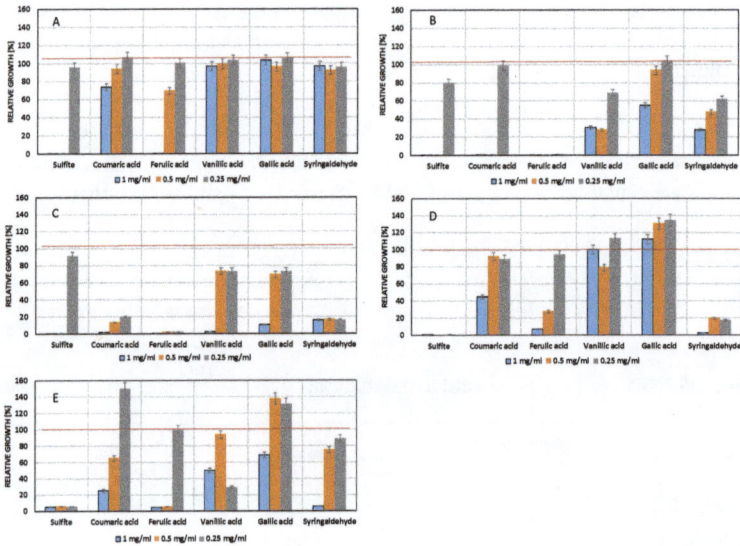

Figure 2. Influence of phenolic compounds on the growth of Saccharomyces cerevisiae 16.1 (**A**), Saccharomyes bayanus HL 77 (**B**); Saccharomyces cerevisiae × S. kudriavzevii × S. bayanus HL 78 (**C**), Wickerhamomyces anomalus AS1 (**D**) and Debaryomyces hansenii 525 (**E**). The red line indicates maximum growth (optical density) of the controls which contained DMSO (5.0 % v/v) but no phenols. Culture medium: Sabouraud-2% Glucose both (pH 3.5).

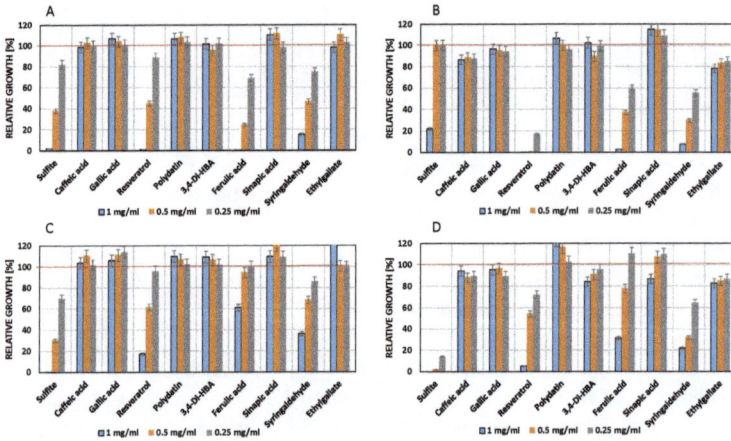

Figure 3. Influence of phenolic compounds on the growth of *S. cerevisiae* 16.1 (**A**) 16.1, *S. bayanus* HL77 (**B**), *W. anomalus* AS1 (**C**) and *D. hansenii* 525 (**D**). The red line indicates maximum growth (optical density) of the controls which contained DMSO (2.5% v/v) but no phenols. Culture medium: Sabouraud-glucose broth (pH 5.5).

3.3. Influence of Phenols on Growth of Lactic Acid Bacteria

The four lactic acid bacteria were completely inhibited by 0.25 mg/mL sulfite (Figure 4). As in the case of yeasts, resveratrol but not its glycosylated form (polydatin) had a strong antimicrobial effect. The cinnamic acids caffeic acid, ferulic acid and sinapic acid decreased cell densities in a concentration-dependent manner. Syringaldehyde at 1.0 mg/mL completely inhibited growth. On the other hand, gallic acid had no effect or stimulated growth of *Pediococcus* and *Oenococcus*.

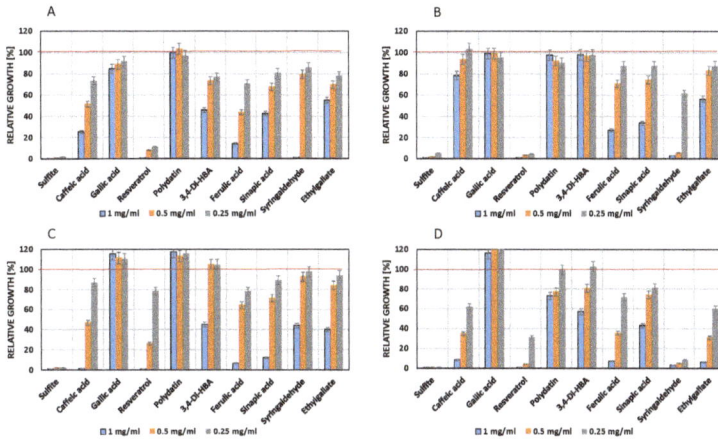

Figure 4. Influence of phenolic compounds on the growth of *Lactobacillus hilgardii* 20176 (**A**), *Lactobacillus plantarum* 20174[T] (**B**), *Pediococcus parvulus* 20332[T] (**C**) and *Oenococcus oeni* 20252 (**D**). The red line indicates maximum growth (optical density) of the controls which contained DMSO (2.5% v/v) but no phenols. Culture medium: MRS (pH 5.5).

3.4. Influence of Phenols on Growth of Acetic Acid Bacteria

In contrast to previous experiments, browning reactions were observed in the cultures, indicating oxidation of the phenols. Thus, only the hitherto most effective compounds were tested (Figure 5). Acetic acid bacteria did not grow in the presence of sulfite. *G. cerinus* 9533 was significantly inhibited by syringaldehyde. The cinnamic acids showed no—and resveratrol only a transient—negative effect. Gallic acid favored growth, as already observed with other microorganisms before. *A. acetii* 358 was inhibited by high concentrations of ferulic acid and resveratrol.

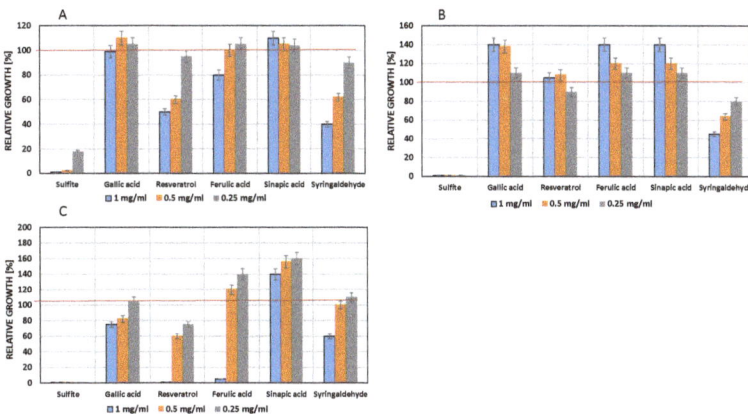

Figure 5. Influence of phenolic compounds on the growth of *Gluconobacter cerinus* 9533 after 24 h (**A**) or 48 h (**B**) and on *Acetobacter acetii* 3508 after 48 h (**C**). The red line indicates maximum growth (optical density) of the controls which contained DMSO (2.5% v/v) but no phenols. Culture medium YPM (pH 6.1) for (**A**) and (**B**) and YPS (pH 6.6) for (**C**).

3.5. Effects of Enzymatic Oxidation

As observed with acetic acid bacteria in our experiments, oxidative transformations may alter the molecular structure and reactivity of phenols. In must and wine, phenols can be enzymatically oxidized by the action of phenoloxidases deriving from grapes or fungal sources [18,20]. In this study, we tested the antimicrobial action of phenolic compounds after oxidation with laccase from *Botrytis cinerea* (Table 2).

Table 2. Effect of *B. cinerea* laccase on the growth effect of phenols [1].

Species	Caffeic Acid	Gallic Acid	Resveratrol	Polydatin	3,4-di-HB	Ferulic Acid	Sinapic Acid	Syring-Aldehyde	Ethyl-gallate
S. cerevisiae	+4.29	−3.98	+2.30	−103.00	+3.94	+5.51	−9.37	+9.63	+4.36
S. bayanus	+24.23	+7.77	±0	−106.00	−3.72	+3.98	−7.00	+2.07	−5.77
D. hansenii	+7.11	−15.58	−4.74	−14.29	−19.94	+1.04	−4.12	−8.75	−32.30
W. anomalus	+9.19	−0.89	+6.17	−88.13	−9.36	−4.98	−7.12	−1.58	−5.37
L. hilgardii	+5.01	+3.71	−8.37	+15.57	+7.99	+1.49	+21.93	+0.49	+3.29
L. plantarum	+10.00	+2.36	−4.41	−11.17	−7.94	+8.77	+23.12	−2.14	+5.06
P. parvulus	+1.51	+8.16	±0	+0.68	+16.01	+8.34	+37.18	−31.65	+29.22
O. oeni	+19.21	+115.74	+9.79	+40.63	+47.38	+12.20	+25.37	+3.01	+30.15

[1] data are given in percentage positive (+) or negative (−) growth compared to the samples without laccase additions; initial concentrations of phenols were 1 mg/mL.

After oxidation with fungal laccase the antimicrobial impact of phenolic compounds was diminished or enhanced compared to the untreated controls. In general, oxidized caffeic acid and ferulic acid were less toxic to all test microorganisms. In addition, lactic acid bacteria took benefit by oxidation of sinapic acid, gallic acid and ethylgallate. In contrast, polydatin inhibited yeast growth much stronger after oxidation than before. No general trend can be derived with respect to laccase oxidation for resveratrol, hydroxybenzoic acid and syringaldehyde.

4. Discussion

Small amounts of SO_2 have a good inhibitory effect on microorganisms which are undesirable in wine production. The wine yeasts *S. cerevisiae*, *S. bayanus* and most of their hybrids have a relatively high insensitivity to this preservative [21]. The use of wine phenols as a SO_2 substitute thus presupposes that the yeasts are not influenced by these compounds in their fermentation activity, and only undesired organisms are inhibited.

In the present work, water-insoluble phenols were dissolved in DMSO. EUCAST (European Committee on Antimicrobial Susceptibility Testing) and CLSI (Clinical and Laboratory Standard Institute) recommend 1% v/v DMSO in the medium in inhibition experiments to exclude any solvent effect. On the other hand, Sardowska-Bartozs et al. [19] found that Saccharomycetes are only inhibited >8% v/v DMSO. We found impaired growth of *S. bayanus* HL77 at 5% v/v DMSO and decided to perform the inhibition experiments at 5% and 2.5% DMSO in the medium. The main results of this work are summarized in Table 3.

4.1. Influence of Phenolic Compounds on Yeast Growth

Examination of the inhibitory effect at culture medium pH 3.5 revealed that non-Saccharomycetes are sensitive to ferulic acid, *p*-coumaric acid and syringaldehyde. In the case of the *Saccharomyces* species, differences in the inhibitory effect were detected. Whereas *S. cerevisiae* 16.1 showed complete inhibition only with 1 mg/mL of ferulic acid or resveratrol, the hybrid HL 78 and *S. bayanus* HL77 revealed marked sensitivities to phenols. Again, p-coumaric acid, ferulic acid and syringaldehyde acted inhibitory. In the experiments conducted at pH 5.5, the negative effect of the cinnamic acids were usually diminished, probably as they are in their deprotonated state (see Section 4.4).

Ferulic acid turned out as the phenolic acid with the strongest inhibitory effect. With *p*-coumaric acid, a good inhibitory effect could also be achieved in some cases. This is consistent with the work of Pastorkova et al. [12], who also observed a high sensitivity of wine-damaging yeasts for hydroxycinnamic acids. Harris et al. [22] ruled out the possible application of ferulic acid as an antifungal agent for the control of *Dekkera* spp.

Table 3. Minimum inhibitory concentration [µg/mL] of phenolic compounds on wine-related microorganisms.

(A) Yeasts [1]					
Compound	*S. cerevisiae*	*S. bayanus*	*S. c.* × *S. k.* × *S. b.*	*W. anomalus*	*D. hansenii*
Caffeic acid	>1000	>1000	nd	>1000	>1000
p-Coumaric acid	>1000	500	1000	1000	>1000
Ferulic acid	1000	250–500	1000	1000	500
Sinapic acid	>1000	>1000	nd	>1000	1000
Gallic acid	>1000	>1000	1000	>1000	>1000
Vanillic acid	>1000	1000	1000	>1000	>1000
Hydroxybenzoic acid	>1000	>1000	nd	>1000	1000
Syringaldehyde	250–1000	250–500	250	250	250–1000
Resveratrol	250	250	250	250–500	250
Polydatin	>1000	>1000	nd	>1000	>1000
Potassium sulfite	250–500	500–1000	500	250	250
(B) Lactic Acid Bacteria					
Compound	*L. hilgardii*	*L. plantarum*	*P. parvulus*	*O. oeni*	
Caffeic acid	250	>1000	250	250	
Ferulic acid	250	250	250	250	
Sinapic acid	250	250	250	250	
Gallic acid	>1000	>1000	>1000	>1000	
Hydroxybenzoic acid	250	>1000	1000	500	
Syringaldehyde	250	250	1000	250	
Resveratrol	250	250	250	250	
Polydatin	>1000	>1000	>1000	>1000	
Potassium sulfite	250	250	250	250	
(C) Acetic Acid Bacteria					
Compound	*G. cerinus*	*A. acetii*			
Ferulic acid	>1000	1000			
Sinapic acid	>1000	>1000			
Gallic acid	>1000	1000			
Syringaldehyde	250	1000			
Resveratrol	>1000	250			
Potassium sulfite	250	250			

[1] results from experimental series 1 and 2; nd: not determined *S. c.* × *S. k.* × *S. b.* = *S. cerevisiae* × *S. kudriavzevii* × *S. bayanus*.

Syringaldehyde gave surprisingly good inhibition results. The compound is one of the less-explored polyphenols and, so far, there is little information on its antimicrobial activity, as well as its other capacities [23]. Studies with the pathogenic yeast *Candida guilliermondii* showed a fungicidal effect of syringaldehyde [24]. In general, it can be stated that the so-called wild yeasts were more strongly inhibited than the Saccharomycetes with syringaldehyde. This observation suggests that it is one of the compounds that could have the potential to be used as SO_2 substitutes. This should be given greater attention in the future.

Resveratrol had an inhibitory effect on most yeast species under investigation including *S. cerevisiae*. This seems in some contrast to studies which revealed that the stilbene increases the live expectancy of this most important wine yeast [9].

4.2. Influence of Phenolic Compounds on the Growth of Lactic Acid Bacteria

Phenolic compounds impaired growth of lactic acid bacteria in a concentration-dependent manner. The inhibition effect increased generally in the order sinapic acid < caffeic acid < p-coumaric acid < ferulic acid < syringaldehyde < resveratrol. Less effective were ethylgallate and 3,4-dihydroxy benzoic acid. Polydatin behaved rather neutral, and gallic acid had even a growth stimulating effect on *P. parvulus* and *O. oeni*. Comparable results have been reported in previous investigations. Stead [25] found that high concentrations of p-coumaric, ferulic and caffeic acids inhibited the growth of *Lactobacillus collinoides* and *Lactobacillus brevis*, whereas low concentrations stimulated their growth. In another study, Rozès and Peres [26] observed that caffeic and ferulic acids negatively affected the growth of *L. plantarum* and increased the proportion of unsaturated fatty acids of the cell membrane. Salih et al. [27] studied the inhibitory effect of hydroxycinnamic acids towards *O. oeni* and *L. plantarum* and found a strong decrease in the growth rate and biomass production in *O. oeni* but only a growth rate decrease was observed in *L. plantarum*. Ferulic acid had the strongest effect, followed by p-coumaric and caffeic acids. Campos et al. [28] reported that hydroxycinnamic acids (and particularly p-coumaric acid) were more inhibitory towards *O. oeni* and *L. hilgardii* than hydroxybenzoic acids; on the other hand, some phenolic acids showed a beneficial effect on growth of *L. hilgardii* [28].

Cinnamic acids can also influence malolactic fermentation of lactic acid bacteria: 50–150 µg/mL caffeic acid supported the degradation of malic acid in Merlot wine, whereas ferulic acid was inhibitory; coumaric acid had even a more negative effect [9].

In line with our results, several authors reported a positive effect of gallic acid on the growth of lactic acid bacteria [29,30]. Alberto et al. [31] found that low concentrations (up to 200 µg/mL) of gallic acid stimulated growth of *L. hilgardii*, whereas at high concentrations (1000 µg/mL) gallic acid was inhibitory to this bacterium. In later works, the same strain of *L. hilgardii* was shown to be able to metabolise gallic acid, producing some phenolic compounds with oxygen-scavenging capacity [32]. Similarly, Theobald et al. [33] reported that epigallocatechine gallate (EGCG) exerts a concentration-dependent impact on the growth of the wine bacterium *Oenococcus oeni* strain B2. It was shown that EGCG had a stimulating effect at 400–500 µg mL^{-1} but an inhibitory effect when present above 540 µg mL^{-1} in the culture media. In relation to this, other authors observed that flavanols with galloyl moiety exhibited more activity on bacteria growth than those without such residues [34].

Landete et al. [35] reported that *L. plantarum* was more sensitive to hydroxycinnamic acids (especially p-coumaric acid) than to hydroxybenzoic acids at the same molar concentrations. García-Ruiz et al. [10] observed that stilbenes had a stronger inhibitory effect towards wine lactic acid bacteria than phenolic acids at the same concentration levels.

4.3. Influence of Phenolic Compounds on the Growth of Acetic Acid Bacteria

Acetic acid bacteria are obligate aerobic microorganisms which are able to oxidize numerous sugars, sugar alcohols and alcohols by the aid of various pyrroloquinoline quinone (PQQ)-dependent membrane-bound dehydrogenases. In wine, they are notorious spoilage microorganisms by the conversion of ethanol into acetic acid [1,36].

At 1 mg/mL syringaldehyde, we found a permanent growth reduction of *A. acetii* 3508 and *G. cerinus* 9533, whereas resveratrol and ferulic acid only inhibited *A. acetii* 3508. The same strain was inhibited by ferulic acid and resveratrol in investigations of Pastorkowa et al. [12]. Syringaldehyde was not included their study. There are only a few other publications dealing with the effect of phenolic compounds on the growth of wine-related acetic acid bacteria. García-Ruiz et al. [11] found that some phenolic plant extracts inhibited growth of *A. aceti* and *Gluconobacter oxydans*.

In our study, we observed that acetic acid bacteria transformed phenols into brownish products. This is in line with a study of Buchert and Niemelä [37], who investigated the effect of some aromatic aldehydes on the growth of *G. oxydans*. The organism detoxified all possible inhibitors except syringaldehyde by simultaneously oxidizing and reducing them to the corresponding acids and alcohols.

4.4. Mode of Antimicrobial Action

Besides differing in structure and complexity, phenolic compounds also exhibit different properties regarding their antimicrobial activities. Benzoic acid was one of the first authorized preservatives for use in the food industry to attain Generally Regarded As Safe (GRAS) status [2]. Although the precise antimicrobial mechanism of phenolic acids is not well known, there is growing evidence that the primary effect is to interfere with the cytoplasmic membrane, increasing its permeability and causing leakage of intracellular constituents such as proteins, nucleic acids, and inorganic ions. They thereby act as decoupling agents leading to a disruption of the transmembrane pH and charge gradient [38]. Phenolic acids are weak organic acids (pKa $\approx$ 4.2) and their antimicrobial activity is considered to be dependent on the concentration of the nondissociated carboxyl group. Due to their lipophilic nature, they can cross the microbial cell membrane in their non-ionic form and acidify the cytosol, causing protein denaturation and interfering with cellular activity [2]. Destruction of the cell wall structure is another bactericidal effect of phenolic compounds as observed for lactic acid bacteria [39]. Gram-negative bacteria are more resistant to phenolic compounds than Gram-positive bacteria due to the presence of their protective outer membrane [2].

Benzaldehydes act primarily by combining with sulfhydryl groups of proteins on the external surface of microorganisms. Other works have shown that these compounds can also cause cellular leakage [2].

The hydroxystilbene resveratrol is one of the best analyzed compounds in wine, especially because of its beneficial influence on human health [40,41]. Being a phytoalexin, the compound is produced by plants in response to biotic and abiotic stress factors and has antifungal and antimicrobial activities. Its glycosylated derivate (polydatin or piceid) occurs in red wines as much as tenfold higher. The antibiotic action mechanism of stilbenes is not well understood. It has been suggested that resveratrol inhibits the respiration of fungal cells, probably by acting as an uncoupling agent or by membrane lipid peroxidation [20]. Pterostilbene is a methoxylated derivate of resveratrol originating from natural sources like wine grapes or blue berries. Its potent capability against methicillin-resistant *Staphylococcus aureus* (MRSA) has been related to bacterial membrane leakage, chaperone protein downregulation, and ribosomal protein upregulation [42].

4.5. Enyzmatic Oxidations

Phenolic compounds of grapes can be rapidly transformed by phenoloxidases, thereby altering their biological properties [18]. Here we found that the antimicrobial action of phenols mostly decreased after laccase oxidation, but also the opposite situation. This may be explained by a different biochemical reactivity of the oxidation products. It is generally believed that fungal laccases as such from *Botrytis cinerea* detoxify phenolic compounds (e.g., the phytoalexin resveratrol) by oxidative transformation to di-, oligo- or polymers which are less bioavailable and bioactive [20]. On the other hand, the first enzymatic oxidation products from polyphenolics are high-reactive quinones which can rapidly react with amino acids, peptides and proteins by Michael-type additions [43]. In the case of enzymes this will lead to their inhibition [9,44].

5. Conclusions

The main subject of this work was to assess the different inhibition behaviors of phenols against wine-associated microorganisms and factors influencing the inhibitory effect, e.g., enzymatic oxidations, pH and organic solvents. The focus was on the evaluation of the individual phenols as a potential sulfite substitute. The most promising compounds were ferulic acid, resveratrol and syringaldehyde. Low concentrations of ferulic acids (250 µg/mL) inhibited potential wine-spoiling bacteria but not growth of wine yeasts. The biotechnological potential of ferulic acid for medical and food applications has already been highlighted [45].

Although antifungal activities of resveratrol are well known, its strong inhibition effect on bacteria and yeasts is somewhat surprising but confirms previous studies [10]. Considering the technological application of resveratrol in the process of wine stabilization, its concentration in wine is too low to effectively inhibit spoilage microorganisms (Table 1). Therefore, wine fortification by addition of pure stilbenes or plant extracts might be considered as prospective strategy [12]. Glycosylated resveratrol (polydatin) was ineffective in our study, whereas another derivate, pterostilbene, displayed up to four times stronger antimicrobial efficacy than resveratrol [12]. Another possibility to increase the trans-resveratrol content in wines is the enzymatic hydrolysis of glycosylated precursors by application of β-glucosidases [46].

A main conclusion which can be drawn from our experiments is that among the phenols tested, syringaldehyde seems to have the highest biotechnological potential, since it acted at relatively low concentrations, especially on the wild yeasts and bacteria. Syringaldehyde as antimicrobial agent has been given little attention so far. This gap should be closed in the future, especially since this phenol also has health-promoting and antioxidant properties [23].

Another point that should be taken into account in the assessment of the inhibitory capacity of phenols are their interactions with each other and with other wine constituents. For example, the combination of different phenols may have a positive or negative effect on the inhibitory effect. Also, the influence of phenoloxidases is not insignificant, as demonstrated in this work. The investigation of these interactions is indispensable in order to be able to make a precise statement about the inhibitory behavior of the phenols. Finally, the influence of antimicrobial plant extracts on the composition and sensorial quality of wines needs to be considered [47]. Particularly, in the case of syringaldehyde the sensory threshold in wine for this compound has been reported at 50 µg/mL, thus at a significant lower level than the lowest concentration (250 µg/mL) tested in this study. The actual minimum antimicrobial concentrations of syringaldehyde will be evaluated in consecutive research.

Acknowledgments: This research project was supported by the Stiftung Rheinland-Pfalz für Innovation (Germany) Project number 961-386261/1051. The authors want to dedicate this article to Helmut König.

Author Contributions: Andrea Sabel, Simone Bredefeld and Martina Schlander performed the experiments and analyzed the data. Harald Claus conceived the study and wrote the paper.

Conflicts of Interest: The authors declare no conflict of interest.

References

1. König, H.; Unden, F.; Fröhlich, J. *Biology of Microorganisms on Grapes, in Must and in Wine*; Springer: Berlin/Heidelberg, Germany, 2009.
2. Campos, F.M.; Couto, J.A.; Hogg, T. Utilisation of natural and by-products to improve wine safety. In *Wine Safety, Consumer Preference, and Human Health*; Moreno-Arribas, M.V., Bartolomé Sualdea, B., Eds.; Springer Int. Publ.: Cham, Switzerland, 2016; pp. 27–49.
3. Pozo-Bayón, M.A.; Monagas, M.; Bartolomé, B.; Moreno-Arribas, M.V. Wine features related to safety and consumer health: An integrated perspective. *Crit. Rev. Food Sci. Nutr.* **2012**, *52*, 31–54. [CrossRef] [PubMed]
4. Papadopoulou, C.; Soulti, K.; Roussis, I.G. Potential antimicrobial activity of red and white wine phenolic extracts against strains of *Staphylococcus aureus*, *Escherichia coli* and *Candida albicans*. *Food Technol. Biotechnol.* **2005**, *43*, 41–46.
5. Cueva, C.; Mingo, S.; Muñoz-González, I.; Bustos, I.; Requena, T.; Del Campo, R.; Martín-Álvarez, P.J.; Bartolomé, B.; Moreno-Arribas, M.V. Antibacterial activity of wine phenolic compounds and oenological extracts against potential respiratory pathogens. *Lett. Appl. Microbiol.* **2012**, *54*, 557–563. [CrossRef] [PubMed]
6. Katalinic, V.; Smole Mozina, S.; Generalic, I.; Skroza, D.; Ljubenkov, I.; Klancik, A. Phenolic profile, antioxidant capacity, and antimicrobial activity of leaf extracts from six *Vitis. vinifera* L. varieties. *Int. J. Food Prop.* **2013**, *16*, 45–60. [CrossRef]
7. Xu, C.; Yagiz, Y.; Zhao, L.; Simonne, A.; Lu, J.; Marshall, M.R. Fruit quality, nutraceutical and antimicrobial properties of 58 muscadine grape varieties (*Vitis. rotindifolia* Michx.) grown in United States. *Food Chem.* **2017**, 149–156. [CrossRef] [PubMed]
8. Landete, J.M. Updated knowledge about polyphenols: Functions, bioavailability, metabolism, and health. *Crit. Rev. Food Sci. Nutr.* **2012**, *52*, 936–948. [CrossRef] [PubMed]
9. Dietrich, H.; Pour-Nikfardjam, M.S. Influence of phenolic compounds and tannins on wine-related microorganisms. In *Biology of Microorganisms on Grapes, in Must and in Wine*; König, H., Unden, F., Fröhlich, J., Eds.; Springer: Berlin/Heidelberg, Germany, 2009; pp. 307–334.

10. García-Ruiz, A.; Moreno-Arribas, M.V.; Martín-Álvarez, P.J.; Bartolomé, B. Comparative study of the inhibitory effects of wine polyphenols on the growth of enological lactic acid bacteria. *Int. J. Food Microbiol.* **2011**, *145*, 426–431. [CrossRef] [PubMed]

11. García-Ruiz, A.; Cueva, C.; Gonzales-Rampinelli, E.M.; Yuste, M.; Torres, M.; Martin-Alvarez, P.J.; Bartolomé, B.; Moreno-Arribas, M.V. Antimicrobial phenolic extracts able to inhibit lactic acid bacteria growth. *Food Control* **2012**, *28*, 212–219. [CrossRef]

12. Pastorkowa, E.; Zakova, T.; Landa, P.; Navakova, J.; Vadlejch, J.; Kokoska, L. Growth inhibitory effect of grape phenolics against wine spoilage yeast and acetic acid bacteria. *Int. J. Food Microb.* **2013**, *161*, 209–213. [CrossRef] [PubMed]

13. Gonzalez-Rompinelli, E.M.; Rodriguez-Bencomo, J.J.; Garcia-Ruiz, A.; Sanchez-Patan, F.; Martin-Alvarez, P.J.; Bartolome, B.; Morena-Arribas, M. A winery-scale trial of the use of antimicrobial plant phenolic extracts as preservatives during wine ageing in barrels. *Food Control* **2013**, *33*, 440–447. [CrossRef]

14. Sabel, A.; Martens, S.; Petri, A.; König, H.; Claus, H. *Wickerhamomyces anomalus* AS1: A new strain with potential to improve wine aroma. *Ann. Microbiol.* **2014**, *64*, 483–491. [CrossRef]

15. Christ, E.; Kowalczyk, M.; Zuchowska, M.; Claus, H.; Löwenstein, R.; Szopinska-Morawska, A.; Renaut, J.; König, H. An exemplary model study for overcoming stuck fermentation during spontaneous fermentation with the aid of a *Saccharomyces* triple hybrid. *J. Agric. Sci.* **2015**, *7*, 18–34. [CrossRef]

16. Zuchowska, M.; Jaenicke, E.; König, H.; Claus, H. Allelic variants of hexose transporter Hxt3p and hexokinases Hxk1p/Hxk2p in strains of *Saccharomyces cerevisiae* and interspecies hybrids. *Yeast* **2015**, *32*, 657–669. [CrossRef] [PubMed]

17. Bäumlisberger, M.; Moellecken, U.; König, H.; Claus, H. The potential of the yeast *Debaryomyces hansenii* H525 to degrade biogenic amines in food. *Microorganisms* **2015**, *3*, 839–850. [CrossRef] [PubMed]

18. Riebel, M.; Sabel, A.; Claus, H.; Xia, N.; Li, H.; König, H.; Decker, H.; Fronk, P. Antioxidant capacity of phenolic compounds on human cell lines as affected by grape-tyrosinase and *Botrytis*-laccase oxidation. *Food Chem.* **2017**, *229*, 779–789. [CrossRef] [PubMed]

19. Sadowska-Bartosz, I.; Pączka, A.; Molon, M.; Bartosz, G. Dimethyl sulfoxid induces oxidative stress in the yeast *Saccharomyces cerevisiae*. *FEMS Yeast Res.* **2013**, *13*, 820–830. [CrossRef] [PubMed]

20. Claus, H.; Sabel, A.; König, H. Wine Phenols and Laccase: An ambivalent relationship. In *Wine Phenolic Composition, Classification and Health Benefits*; El Rayess, E.Y., Ed.; Nova publishers: New York, NY, USA, 2014; pp. 155–185.

21. Dittrich, H.H.; Großmann, M. *Mikrobiologie des Weines 4. Auflage*; Verlag Eugen Ulmer: Stuttgart, Germany, 2011.

22. Harris, V.; Jiranek, V.; Ford, C.M.; Grbin, P.R. Inhibitory effect of hydroxycinnamic acids on *Dekkera* spp. *Appl. Microbiol. Biotechnol.* **2010**, *86*, 721–729. [CrossRef] [PubMed]

23. Ibrahim, M.N.M.; Sriprasanthi, R.B.; Shamsudeen, S.; Adam, F.; Bhawani, S.A. A concise review of the natural existence, synthesis, properties and applications of syringaldehyde. *BioResources* **2012**, *7*, 4377–4399.

24. Kelly, C.; Jones, O.; Barnhart, C.; Lajoie, C. Effect of furfural, vanillin and syringaldehyde on *Candida guilliermondii*: Growth and xylitol biosynthesis. *Appl. Biochem. Biotechnol.* **2008**, *148*, 97–108. [CrossRef] [PubMed]

25. Stead, D. The effect of hydroxycinnamic acids on the growth of wine-spoilage lactic acid bacteria. *J. Appl. Bact.* **1993**, *75*, 135–141. [CrossRef]

26. Rozès, N.; Peres, C. Effects of phenolic compounds on the growth and fatty acid composition of *Lactobacillus plantarum*. *Appl. Microbiol. Biotechnol.* **1998**, *49*, 108–111. [CrossRef]

27. Salih, A.G.; Le Quéré, J.M.; Drilleau, J.F. Effect of hydroxycinnamic acids on the growth of lactic bacteria. *Sci. Aliment.* **2000**, *20*, 537–560. [CrossRef]

28. Campos, F.M.; Couto, J.A.; Hogg, T. Influence of phenolic acids on growth and inactivation of *Oenococcus oeni* and *Lactobacillus hilgardii*. *J. Appl. Microbiol.* **2003**, *94*, 167–174. [CrossRef] [PubMed]

29. Stead, D. The effect of chlorogenic, gallic and quinic acids on the growth of spoilage strains of *Lactobacillus collinoides* and *Lactobacillus brevis*. *Lett. Appl. Microbiol.* **1994**, 112–114. [CrossRef]

30. Vivas, N.; Lonvaud-Funel, A.; Glories, Y. Effect of phenolic acids and anthoyanins on growth, viability and malolactic activity of a lactic acid bacterium. *Food Microbiol.* **1997**, 291–299. [CrossRef]

31. Alberto, M.R.; Farías, M.E.; Manca de Nadra, M.C. Effect of gallic acid and catechin on *Lactobacillus hilgardii* 5w growth and metabolism of organic compounds. *J. Agric. Food Chem.* **2001**, *49*, 4359–4363. [CrossRef] [PubMed]

32. Alberto, M.R.; Gómez-Cordovés, C.; Manca de Nadra, M.C. Metabolism of gallic acid and catechin on *Lactobacillus hilgardii* from wine. *J. Agric. Food Chem.* **2004**, *52*, 6465–6469. [CrossRef] [PubMed]

33. Theobald, S.; Pfeiffer, P.; Zuber, U.; König, H. Influence of epigallocatchin gallate and phenolic compounds from green tea on the growth of *Oenococcus oeni*. *J. Appl. Microbiol.* **2008**, *104*, 566–672. [CrossRef] [PubMed]

34. González de Llano, D.; Gil-Sánchez, I.; Esteban-Fernández, A.; Ramos, A.M.; Cueva, C.; Moreno-Arribas, M.V.; Bartolomé, B. Some contributions to the study of oenological lactic acid bacteria through their interaction with polyphenols. *Beverages* **2016**, *2*, 27. [CrossRef]

35. Landete, J.M.; Rodríguez, H.; de las Rivas, B.; Munoz, R. High-added-value antioxidants obtained from the degradation of wine phenolic by *Lactobacillus plantarum*. *J. Food. Prot.* **2007**, *70*, 2670–2675. [CrossRef] [PubMed]

36. Guillamón, J.M.; Mas, A. Acetic acid bacteria. In *Molecular Wine Microbiology*; Carrascosa, A.V., Muñoz, R., González, R., Eds.; Elsevier: London, UK, 2011; pp. 227–255.

37. Buchert, J.; Niemelä, K. Oxidative detoxification of wood-derived inhibitors by *Gluconobacter oxydans*. *J. Biotechnol.* **1991**, *18*, 1–12. [CrossRef]

38. Amborabé, B.E.; Fleurat-Lessard, P.; Chollet, J.F.; Roblin, G. Antifungal effects of salicylic acid and other benzoic derivates towards *Eutypa lata*: Structure-activity relationship. *Plant Phys. Biochem.* **2002**, *40*, 1051–1060. [CrossRef]

39. Rodríguez, H.; Curiel, J.A.; Landete, J.M.; de las Rivas, B.; de Filipe, F.L.; Gómez-Cordovés, C.; Mancheño, J.M.; Muñoz, R. Food phenolics and lactic acid bacteria. *Int. J. Food. Microbiol.* **2009**, *132*, 79–90. [CrossRef] [PubMed]

40. Suart, J.A.; Robb, E.A. *Bioactive Polyphenols from Wine Grapes*; Springer: New York, NY, USA, 2013.

41. El Rayess, Y. *Wine–Phenolic Composition, Classification and Health Benefits*; Nova Publishers: New York, NY, USA, 2014.

42. Yang, S.C.; Tseng, C.H.; Wang, P.W.; Lu, P.L.; Weng, Y.H.; Yen, F.L.; Fang, J.Y. Pterostilbene, a methoxylated resveratrol derivate, efficiently eradicates planctonic, biofilm, and intracellular MRSA by topical application. *Front. Microbiol.* **2017**, *8*, 1103. [CrossRef]

43. Fernandes, M.S.; Kerkar, S. Microorganisms as a source of tyrosinase inhibitors: A review. *Ann. Microbiol.* **2017**, *67*, 343–358. [CrossRef]

44. Pourcel, L.; Routaboul, J.M.; Cheynier, V.; Lepiniec, L.; Debeaujon, I. Flavanoid oxidation in plants: From biochemical properties to physiological functions. *Trends Plant Sci.* **2006**, *12*, 29–36. [CrossRef] [PubMed]

45. Kumar, N.; Pruthi, V. Potential applications of ferulic acid from natural sources. *Biotechnol. Rep.* **2014**, *4*, 86–93. [CrossRef] [PubMed]

46. Schwentke, J.; Sabel, A.; Petri, A.; König, H.; Claus, H. The yeast *Wickerhamomyces. anomalus* AS1 secretes a multifunctional exo-β-1,3-glucanase with implications for winemaking. *Yeast* **2014**, *31*, 349–359. [CrossRef] [PubMed]

47. García-Ruiz, A.; Rodríguez-Bencomo, J.J.; Garrido, I.; Martín-Álvarez, M.; Moreno-Arribas, M.V.; Bartolomé, B. Assessment of the impact of the additions of antimicrobial plant extracts to wine: Volatile and phenolic composition. *J. Sci. Food Agric.* **2013**, *93*, 2507–2516. [CrossRef] [PubMed]

beverages

MDPI

Article

Characterization of an Antioxidant-Enriched Beverage from Grape Musts and Extracts of Winery and Grapevine By-Products

Tabita Aguilar [1], Johannes de Bruijn [1,*], Cristina Loyola [1], Luis Bustamante [2], Carola Vergara [2], Dietrich von Baer [2], Claudia Mardones [2] and Ignacio Serra [3]

[1]   Department of Agroindustry, University of Concepcion, Av. Vicente Mendez 595, Chillan 3780000, Chile; tabbyfebbe@gmail.com (T.A.); cloyola@udec.cl (C.L.)

[2]   Department of Instrumental Analysis, University of Concepcion, Barrio Universitario s/n, Concepcion 4030000, Chile; lbustamante@udec.cl (L.B.); carolavergara@udec.cl (C.V.); dvonbaer@udec.cl (D.v.B.); cmardone@udec.cl (C.M.)

[3]   Department of Vegetal Production, University of Concepcion, Av. Vicente Mendez 595, Chillan 3780000, Chile; iserra@udec.cl

Received: 18 October 2017; Accepted: 28 December 2017; Published: 8 January 2018

Abstract: The recovery of antioxidants from complex winery and grapevine by-products into *Vitis vinifera* must offers new opportunities for wine grapes by the development of a new, enriched fruit juice. However, this demands the search for new valorization methods to get hold of additional antioxidant compounds. The objective of this study was to find a novel functionality for grape pomace, grapevine leaves, and canes by its reuse as a functional matrix for the extraction of antioxidants into grape must. After thermomaceration, 22 polyphenols were identified by high performance liquid chromatography and mass spectrometry. Grape pomace was a good source of anthocyanins (malvidin-3-glucoside), while flavonols (quercetin-3-hexoside) and phenolic acids (caftaric acid) were the main phenolic compounds in leaf extracts. Catechin dimer was the only polyphenol compound present in all of the matrices. Enriched grape juice comprised by 40:20:40 ($v/v/v$) of pomace, leaf, and cane extracts, yielded an oxygen radical absorbance capacity of pirogallol red and fluorescein ratio of 0.70, indicating that the reactivity of antioxidants present in enriched grape juice was at least as efficient as other polyphenol-rich beverages. Thus, pomace, leaves and canes supply additional polyphenols to grape must that results into a beverage with promissory antioxidant activity and potential health benefits.

Keywords: grape juice; thermomaceration; antioxidants; polyphenols; *Vitis vinifera*

1. Introduction

Since the observation of a lower mortality rate of coronary heart disease in France when compared to Northern European countries, known as the "French paradox" [1], a number of studies showed the health-promoting effects of phenolic compounds that are present in grapes and grape-derived products, including pure grape juice [2–5]. Consequently, there is a steady global rise of grape juice production over the last thirty years to fulfil a growing demand for pure grape juice by health-conscious consumers. However, even if grape juice meets all health requirements, flavor and other product attributes are critical for consumer acceptance. Thus, the increasing demand for healthy, sensory attractive fruit juices by more demanding, better-educated consumers requires a continuous need for the development of new juice products.

After comparing 13 commercially available fruit juices and juice drinks, purple grape juice contained the highest levels of polyphenols and antioxidants [6]. Phenolic acids and flavan-3-ols were the predominant compounds in white grape juice, while the major groups of polyphenols that were found in purple and red grape juices comprised anthocyanins and flavan-3-ols [6–9]. In particular, several health benefits are associated with the consumption of purple grape juice, such as an improved endothelial function, protection against LDL cholesterol oxidation, decrease in LDL-HDL cholesterol ratio, inhibition of atherosclerosis, improved neurocognitive function, and improved antioxidant biomarkers in blood [2,10–12]. The main beneficial effects of purple grape juices may be due to their contents of flavonoids. In particular, procyanidin dimers, flavonols, and flavan-3-ols show high antioxidant capacity among other polyphenols [13]. Primarily, flavonols and proanthocyanins are associated with a marked decrease in platelet superoxide production and inhibition of platelet aggregation [14], while oligomeric procyanidins improve vascular health [15]. Moreover, anthocyanins, catechin, procyanidins, and E-resveratrol from grape skins and seeds show an inhibition of the growth of human cancer cells [16–18], while anthocyanins and E-resveratrol may also suppress inflammatory reactions [19].

Despite of high phenolic contents of grapes, the processing of grapes results in high amounts of by-products, whereby a major part of the phenolics remain within the grape pomace after processing. However, the content in bioactive phytochemicals that are detected in grape residues shows a strong variation due to different agro-climatic and recovery conditions. Concerning the importance of the up to nine million tons of vine and winery by-products produced globally every year from grape industrialization [20], these by-products could provide extra desirable ingredients for health-food applications improving juice quality [21,22]. Therefore, the aim of the present study was to find a novel functionality for grape pomace, grapevine leaves, and canes by its reuse as a functional matrix for the extraction of antioxidants into must of *Vitis vinifera* grapes. At the same time, it helps to add value to those minor red grape cultivars that are not destined to the production of fine wines.

2. Materials and Methods

2.1. Reagents and Standards

Commercial standards of delphinidin-3-glucoside, petunidin-3-glucoside, malvidin-3-glucoside, cyanidin-3-glucoside, quercetin-3-glucuronide, and chlorogenic acid were obtained from PhytoLab (Vestenbergsgreuth, Germany). Extrasynthese (Lyon, France) provided peonidin-3-glucoside. Commercial standards of quercetin-3-rutinoside, quercetin-3-glucoside, gallic acid, ferulic acid, *p*-coumaric acid, (+)-catechin, (−)-epicatechin and Trolox (6-hydroxy-2,5,7,8-tetramethylchromane-2-carboxylic acid) were purchased from Sigma–Aldrich (St. Louis, MO, USA). Merck (Darmstadt, Germany) provided formic acid, acetonitrile, methanol and water (all HPLC grade), while neocuproine hemihydrate was obtained from Fluka (Buchs, Switzerland).

2.2. Samples

Grapes, grapevine leaves and canes from *Vitis vinifera* L., cvs. País (PA; an ancient, red cultivar) and Lachryma Christi (LC; a Teinturier cultivar, commonly used for blending with pale red wines to intensify red color) were collected from two vineyards that were located in the Itata Valley, San Nicolas, Chile (36°33′ S–72°10′ W and 36°30′ S–72°05′ W, respectively) between March and June 2013. After destemming, grapes were crushed (PAS.0540, Bertuzzi, Brugherio—Milano, Italy) and pressed (D.64625, Willmes, Bensheim, Germany), prior to the collection of grape must and pomace (skins and seeds) and storage at −20 °C. After vintage, autumn leaves and canes were cut and stored at −20 °C until solid-liquid extraction.

2.3. Extraction Procedure

Aqueous extracts from unfrozen, ground (1–2 mm) pomace, leaves and canes were prepared by using grape must as a solvent. Thermomaceration was carried out by mixing 55 g of winery or grapevine by-product with 445 g of grape must in a 1-L double-wall glass vessel, stirred at

500 rpm (Barnstead Thermolyne, Super Nuova magnetic stirrer, Thermo Scientific, Ashville, NC, USA). Temperature was set at 60 °C using a Thermo/Haake DC10-K10 circulating heating bath (Haake, Karlsruhe, Germany), while the extraction vessel was duly covered by parafilm and aluminum foil to avoid solvent loss and light influence. The process time was fixed at 4, 6, and 8 h for cane, pomace, and leaf samples, respectively, in order to achieve a grape must with a maximum amount of polyphenols and antioxidants [23]. Then, enriched grape juice was prepared by mixing pomace, leaf and cane extracts as follows: after filtration using 20–25 μm nylon filter bags, pomace, leaf, and cane extracts were mixed in a ratio of 40:20:40 (*v/v/v*). After combining 250 mL of the País mixture to an equal amount of Lachryma Christi mixture, the resulting blend, called enriched grape juice, was bottled, heated (63 °C for 30 min), and stored at 20 °C until analysis.

2.4. Analytical Methods

2.4.1. Physicochemical Characterization

Standard enological parameters, such as pH, total acidity, soluble solids, free and total SO_2, and chromatic characteristics, were measured in must and enriched juice samples, according to the official OIV methods of analysis [24]. All of the measurements were done in triplicate.

2.4.2. Spectrophotometric Assays

Spectrophotometric analyses were performed using an Analytik Jena Specord 200 Plus spectrophotometer (Jena, Germany) set at the appropriate wavelength for each assay.

Total polyphenol content was measured by using the Folin-Ciocalteu colorimetric assay and the results were reported in mg/kg of gallic acid equivalents [25]. Monomeric anthocyanin content was analyzed using the pH-differential method, expressing the results in mg/kg of malvidin-3-glucoside equivalents [26]. Total flavonoid content was evaluated using a colorimetric assay with aluminum chloride using catechin as standard [27].

Additionally, total antioxidant capacities were evaluated by the ABTS (2,2'-azinobis(3-ethylbenzothiazoline -6-sulfonic acid)) radical cation assay, CUPRAC (cupric reducing antioxidant capacity) method, ORAC (oxygen radical absorbance capacity)—fluorescein (FL), and ORAC—pirogallol red (PGR) assays, as described previously [23,28–30]. In all of the assays, Trolox was used as reference compound and results were expressed in terms of mmol Trolox equivalent antioxidant capacity per kg of sample. All of the measurements were done in triplicate.

2.4.3. Sample Preparation

Prior to HPLC separation, a solid-phase extraction using Oasis MCX cartridges (Waters, Milford, MA, USA) containing a mixture of reverse-phase and cation exchange materials allowed for the separation of anthocyanins from the remaining phenolic compounds to allow their determination without interference. Five milliliters of each must or extract sample were diluted with 5 mL of 0.1 M hydrochloric acid. This solution was passed through the 500 mg Oasis MCX cartridge previously conditioned with 5 mL of methanol and 5 mL of water. After rinsing with 5 mL of 0.1 M hydrochloric acid and 5 mL of water, the fraction containing flavonols, flavan-3-ols, and phenolic acid derivatives was eluted by passing 3 × 5 mL of methanol, while the anthocyanins remained in the solid phase. The anthocyanins were recovered by eluting 10 mL of 5% *w/v* ammonium hydroxide in methanol. Subsequently, the solvents of these fractions were removed by vacuum rotary evaporation and the residues dissolved in 5 mL of mobile phase used in HPLC separation.

2.4.4. HPLC-DAD-ESI-MS/MS

HPLC separation, identification, and quantification of specific phenolic compounds were carried out using a Shimadzu HPLC Nexera system (Kyoto, Japan). This equipment consists of a quaternary LC-30AD pump, DGU-20A$_{5R}$ degasser unit, CTO-20AC oven, SIL-30AC auto-sampler, CBM-20A controller system, and UV-Vis diode array spectrophotometer (model SPD-M20A), coupled in tandem with a QTrap LC/MS/MS 3200 Applied Biosystems MDS Sciex system (Foster City, CA, USA). The detector offers wide linearity (2.5 AU) and a noise level of 0.6×10^{-5} AU for a wavelength from 190 to 700 nm. Instrument control and data collection were done using CLASS-VP DAD Shimadzu Chromatography Data System and Analyst software (version 1.5.2) for MS/MS analysis.

Anthocyanins were separated by HPLC using a C18 YMC 5 μm, 250×4.6 mm column with a C18 Nova-Pak 4 μm, 22×3.9 mm precolumn (Waters, Milford, MA, USA) with a flow rate of 0.3 mL/min at 30 °C. The injection volume was 50 μL. The mobile phase consisted of 0.1% v/v trifluoroacetic acid in water (A) and 100% acetonitrile (B). The gradient program was from 10% to 20% of solvent B in 15 min, followed by 6 min of stabilization, from 20% to 27% in 5 min, followed by 10 min stabilization, from 27% to 100% in 1 min, and from 100% to 10% in 1 min, followed by an isocratic step of 10 min at 10% B.

HPLC separation of flavonols, flavan-3-ols and phenolic acid derivatives was carried out using a Kinetex C18 column (core shell, 150×4.6 mm, 2.6 μm) with a SecurityGuard AJO-8768 C18 cartridge (Phenomenex, Torrance, CA, USA). The injection volume was 10 μL. A binary mobile phase of 0.1% v/v formic acid in water and acetonitrile was used at a flow rate of 0.5 mL/min. The acetonitrile gradient ranged from 15% to 25% acetonitrile for 14 min, from 25% to 35% for 11 min, from 35% to 100% for 1 min, from 100% to 15% for 1 min, followed by a stabilization period of 10 min. The column temperature was set at 30 °C for flavan-3-ols and phenolic acid derivatives, and at 40 °C for flavonols.

The analyses of stilbenoids were carried out using a C18 Kromasil 5 μm, 250×4.6 mm column (Akzo Nobel, Bohus, Sweden) with a C18 Nova-Pak Waters 22×3.9 mm, 4 μm precolumn (Waters, Milford, MA, USA) at 30 °C, using a mobile phase gradient consisting of 0.1% v/v formic acid in water (solvent A) and acetonitrile (solvent B). The injection volume was 25 μL. The flow rate was 0.5 mL/min, and the gradient program was from 15% to 20% of solvent B in 5 min, 20% to 44.5% in 45 min, and 44.5% to 100% in 1 min, followed by an isocratic step of 9 min at 100% and stabilization for 5 min at 15% of B.

The identity of phenolic compounds was assigned by ESI-MS/MS setting the following parameters: negative ionization mode; collision energy, 5 V; ionization voltage, -4000 V; capillary temperature, 450 °C; nebulizer gas, 15 psi. For identification of anthocyanins, a positive ionization mode was used. The identity assignation of compounds was done by comparison of their retention time (t_R), UV-Vis spectra and mass (MS/MS) spectra with those of their respective commercially available standards. Quantification was performed using a DAD chromatogram extracted at 280 nm for flavan-3-ols, 306 nm for stilbenoids, 320 nm for phenolic acid derivatives, 360 nm for flavonols, and 518 nm for anthocyanins. For quantitative determinations, calibration curves were made with the commercial standards for flavan-3-ols, stilbenoids, phenolic acids, flavonols, and anthocyanins. Standard solutions spanning the concentration range from 1.0 to 80 mg/L were prepared by appropriate dilution of standard solutions in solvent A. The limits of detection and quantification were three and ten times the noise signal from the chromatograms of low standard concentration.

2.5. Sensory Analysis

Sensory analysis was performed using a tasting panel of twelve trained judges (six men and six women, aged 25–62 years, consisting of students and university employees) that compared sensory attributes of several mixtures of pomace, leaf, and cane extracts of both País and Lachryma Christi grapes. Panelists were seated in separate booths, each with appropriate lighting, ventilation, and free from noise and other distracting stimuli. Thirty-milliliter samples coded with a random three-digit number were presented at 20 °C to each panelist in blue color glass cups. Samples were tested in three sessions over different days by preference ranking to evaluate flavor, taste, and tactile attributes, where the judges were presented with the samples and were instructed to indicate their preference for each sample on a hedonic scale (0–10) and to order them from least (0 score) to most preferred ones (10 score). Panelists had to eat some piece of cracker and rinse their mouth with water to reduce carry-over effects between sample evaluations. At the end of each session, the judges completed a questionnaire to give their perceptions of sensory attributes. The questionnaire included a list of descriptors of odor (fruity, herbal, flowery, woody, spicy), taste (sweetness, acidity, bitterness, salty), and tactile attributes (astringency, sandy).

2.6. Statistical Analysis

Analysis of variance was performed to assess statistically significant differences between data of grape juice samples at a confidence level of 95% using the Statgraphics Centurion XVII software, version 17.1.06 (Statpoint Technologies, Warrenton, VA, USA). Differences on the means were assessed by Duncan's multiple range test at a significance level of $p < 0.05$, using the Fisher's least significant difference procedure. Preference ranking data were analyzed using the Kruskal-Wallis test for non-homogeneous samples at $p = 0.05$.

3. Results and Discussion

3.1. Phenolic Composition

In the current study, 22 polyphenol compounds, including five anthocyanins, eight flavonols, six phenolic acids, and three flavan-3-ols, were identified by HPLC-DAD-ESI-MS/MS in grape musts and extracts (Table 1). An unknown compound from Lachryma Christi grapes being a hydroxycinnamic acid derivative according to its UV-spectra did not provide enough molecular information in order to establish its identity. Malvidin-3-glucoside, quercetin-3-hexoside, caftaric acid, and catechin dimers were the predominant compounds that were detected in these samples. Amongst polyphenols, catechin dimer was the only component detected both in PA and LC must. Moreover, this component was detected in all of the samples. A low number of phenolic acids and flavan-3-ols, and the absence of anthocyanins and flavonols are observed in PA must, in contrast to LC must, which was particularly rich in flavan-3-ols and anthocyanins. This agrees with the high levels of flavan-3-ols and anthocyanins reported for purple grape juice among 13 commercial fruit juices [6].

Pomace of País grapes is an additional source of a diversity of phenolic compounds, supplying substantial amounts of malvidin-3-glucoside, quercetin-3-hexoside, caftaric and coumaric acid, catechin, and epicatechin to grape must (Table 2). Pomace from Lachryma Christi grapes yielded an extract, which was highly fortified in anthocyanins with a maximum concentration of 1563 mg malvidin-3-glucoside/kg extract (Table 2).

Additionally, delphinidin-3-glucoside and petunidin-3-glucoside are localized typically in grape skins and seeds, but not in the pulp, being absent in must from Lachryma Christi grapes. Anthocyanins from grape skins include malvidin-3-glucoside, delphinidin-3-glucoside, peonidin-3-glucoside, petunidin-3-glucoside, amongst others [31,32]. However, grape seeds may also provide anthocyanins to the must that depends on process conditions, such as time, temperature, and solvent and ultrasound assistance to extraction [33]. Malvidin-acetyl-glucoside, detected in must and pomace extract from Lachryma Christi grapes seems to be involved in complex formation with other organic cofactors, which results in a product with a deep dark purple color. In Pinot noir and Sangiovese cultivars, the lack of acetylated anthocyanins caused a minimum level of co-pigmentation [34].

Furthermore, pomace of Lachryma Christi cultivar provided other phenolic compounds, such as myricetin, quercetin, and isorhamnetin derivatives, and hydroxybenzoic and hydroxycinnamic acid derivatives, lacking in grape must. Although most of the anthocyanins and flavan-3-ols (catechin, epicatechin, and procyanidin B2) are generally removed as skins and seeds during juice processing [35–37], bioactive contents from grape pomace may still become available to fortify juice after applying thermomaceration.

Table 1. Chemical characterization (HPLC-DAD-ESI-MS/MS) of polyphenols in grape musts and extracts.

Name	t_R (min)	[M-H] (m/z)	Product Ions (m/z)	λ_{max} (nm)	Detected in [1]
Anthocyanins					
Delphinidin-3-glucoside	16.86	465	303	524, 277, 343	f
Petunidin-3-glucoside	20.23	479	317, 302	525, 277	f
Peonidin-3-glucoside	22.54	463	301, 286	517, 279	e, f, g, h
Malvidin-3-glucoside	23.36	493	331, 315, 287	527, 277, 346	b, e, f, g, h
Malvidin-acetyl-glucoside	34.30	535	331, 315, 287	529, 525	e,f
Flavonols					
Myricetin-3-hexoside	7.04	493	317, 179, 299, 151, 271	556	c, f, g, h
Quercetin-3-rutinoside	8.02	609	301, 271, 256, 279, 151	354	c, g
Quercetin-3-hexoside	8.75	463	300, 271, 255, 179, 151	357	b, c, f, g, h
Quercetin-3-glucuronide	8.93	477	301, 151, 179, 274, 283	354	c, f, g, h
Kaempferol-3-hexoside	10.70	447,5	284, 205, 227, 183, 135, 197	346	c, g
Kaempferol-3-glucoside	11.70	447,3	284, 255, 227, 153, 179, 241	346	c, g
Isorhamnetin-3-hexoside	12.60	477	315, 285, 271, 299, 243, 151, 179	354	f, g
Isorhamnetin-3-glucuronide	13.30	491	315, 300, 271, 255, 179, 151	353	g
Phenolic acids					
Gallic acid hexoside	6.76	331	271, 211, 169, 151, 125	276	a, c, d, f, g, h
Protocatechuic acid hexoside	7.38	315	153, 123	278	c, d, f, g, h
Ferulic acid hexoside	8.09	355	193, 165	275	c, f, g
Chlorogenic acid	8.55	353	191, 179, 161, 135	320	a, g
Caftaric acid	10.17	311	179, 149, 135	328, 300(sh) [2]	a, b, c, d, f, g, h
p-Coumaric acid	13.47	295	163, 149, 119	311, 300(sh) [2]	b, c, d, f, g, h
Flavan-3-ols					
Catechin dimer	9.01	577	451, 425, 407, 289	280	a, b, c, d, e, f, g, h
(+)-Catechin	11.93	289	245, 203, 179, 161, 125, 137	280	b, c, d, e, f, g, h
(−)-Epicatechin	14.40	289	245, 203, 203, 179, 151, 137, 123, 109	279	b, c, d, e, f, g, h

[1] (a) País must, (b) País pomace extract, (c) País leaf extract, (d) País cane extract, (e) Lachryma Christi must, (f) Lachryma Christi pomace extract, (g) Lachryma Christi leaf extract, (h) Lachryma Christi cane extract. [2] Sh: shoulder.

Table 2. Concentration of phenolic compounds in musts and extracts from País and Lachryma Christi grapes [1].

Name	Must PA	Must LC	Pomace Extract PA	Pomace Extract LC	Leaf Extract PA	Leaf Extract LC	Cane Extract PA	Cane Extract LC
Anthocyanins								
Delphinidin-3-glucoside	n.d.	n.d.	n.d.	232 ± 0	n.d.	n.d.	n.d.	n.d.
Petunidin-3-glucoside	n.d.	n.d.	n.d.	272 ± 1	n.d.	n.d.	n.d.	n.d.
Peonidin-3-glucoside	n.d.	527 ± 1 b	n.d.	533 ± 2 a	n.d.	426 ± 1 c	n.d.	235 ± 1 d
Malvidin-3-glucoside	n.d.	1419 ± 1 b	97.5 ± 0.0 e	1563 ± 0 a	n.d.	717 ± 0 c	n.d.	638 ± 1 d
Malvidin-acetyl-glucoside	n.d.	348 ± 1 b	n.d.	396 ± 0 a	n.d.	n.d.	n.d.	n.d.
Flavonols								
Myricetin-3-hexoside	n.d.	n.d.	n.d.	34.4 ± 0.2 b	14.1 ± 0.9 c	59.0 ± 0.3 a	n.d.	8.0 ± 0.0 d
Quercetin-3-rutinoside	n.d.	n.d.	n.d.	n.d.	9.2 ± 0.4 b	22.1 ± 0.4 a	n.d.	n.d.
Quercetin-3-hexoside	n.d.	n.d.	21.4 ± 0.4 c	38.4 ± 0.7 c	834 ± 41 b	1100 ± 14 a	n.d.	5.9 ± 0.2 c
Quercetin-3-glucuronide	n.d.	n.d.	n.d.	43.6 ± 0.6 c	204 ± 0 b	568 ± 3 a	n.d.	5.3 ± 0.6 d
Kaempferol-3-hexoside	n.d.	n.d.	n.d.	n.d.	11.7 ± 0.9 b	19.4 ± 0.1 a	n.d.	n.d.
Kaempferol-3-glucoside	n.d.	n.d.	n.d.	n.d.	69.1 ± 5.7 b	78.3 ± 0.2 a	n.d.	n.d.
Isorhamnetin-3-hexoside	n.d.	n.d.	n.d.	15.8 ± 0.0 b	n.d.	35.0 ± 0.0 a	n.d.	n.d.
Isorhamnetin-3-glucuronide	n.d.	n.d.	n.d.	n.d.	n.d.	5.8 ± 0.0	n.d.	n.d.
Phenolic acids								
Gallic acid hexoside	1.7 ± 0.0 d	n.d.	n.d.	2.1 ± 0.1 c	4.5 ± 0.0 b	6.3 ± 0.0 a	1.5 ± 0.0 e	1.7 ± 0.0 d
Protocatechuic acid hexoside	n.d.	n.d.	n.d.	2.0 ± 0.0 c	4.6 ± 0.1 b	6.6 ± 0.0 a	1.5 ± 0.0 d	2.0 ± 0.2 c
Ferulic acid hexoside	n.d.	n.d.	n.d.	4.2 ± 0.0 a	3.9 ± 0.0 b	3.3 ± 0.0 c	n.d.	n.d.
Chlorogenic acid	1.7 ± 0.0 b	n.d.	n.d.	n.d.	n.d.	10.4 ± 0.1 a	n.d.	n.d.
Caftaric acid	4.0 ± 1.6 e	n.d.	36.6 ± 1.0 c	4.0 ± 0.0 e	76.8 ± 1.4 b	125 ± 0 a	16.2 ± 1.3 d	5.3 ± 0.7 e
p-Coumaric acid	n.d.	n.d.	8.3 ± 0.4 c	2.6 ± 0.0 e	17.5 ± 0.1 b	20.9 ± 0.0 a	6.2 ± 0.0 d	2.2 ± 0.2 f
Flavan-3-ols								
Catechin dimer	7.8 ± 1.0 c	5.9 ± 0.9 c	10.9 ± 2.5 c	26.7 ± 0.3 b	22.2 ± 0.1 b	54.1 ± 0.3 a	2.9 ± 0.2 c	24.0 ± 1.4 b
(+)-Catechin	n.d.	54.2 ± 0.2 c	42.3 ± 0.1 f	96.6 ± 0.3 b	48.5 ± 0.2 d	110 ± 0 a	6.8 ± 0.2 g	45.5 ± 1.2 e
(−)-Epicatechin	n.d.	14.4 ± 0.0 e	25.5 ± 0.0 d	92.4 ± 0.5 a	36.1 ± 0.9 c	82.4 ± 1.0 b	8.0 ± 0.0 f	11.6 ± 0.9 e

[1] Data are expressed as mean ± standard deviation in mg/kg wet weight (n = 2). Different letters in the same row indicates statistically significant difference ($p < 0.05$). PA: cultivar País, LC: cultivar Lachryma Christi.

Vine leaves are able to enrich grape musts in flavonols, phenolic acids, and flavan-3-ols (Table 2). However, the availability of flavonols, flavan-3-ols and other flavonoids in grapevine parts depends on grape variety, stage of maturation, and recollection [38]. Concerning the flavonols, quercetin-3-hexoside was the most abundant species followed by quercetin-3-glucuronide, whereas myricetin and kaempferol derivatives were found at lower levels. Moreover, quercetin-3-rutinoside, kaempferol-3-hexoside, kaempferol-3-glucoside, and isorhamnetin-3-glucuronide were exclusively found in leaf extracts, being categorized as leaf-associated components. However, kaempferol-3-glucoside has been detected before in very low concentrations (0.001 mg/g fresh weight) in *V. vinifera* grapes [39]. Furthermore, leaves did not supply additional anthocyanins to grape musts. Additionally, thermomaceration provokes a significant loss of anthocyanins (derivatives of peonidin and malvidin) in LC leaf extracts. The concentration of anthocyanins retained in leaf extracts depends on a combination of thermo-induced effects, including co-pigmentation with flavonols, condensation and polymerization with flavan-3-ols, partitioning between leaves and must, and adsorption to the solid phase [34].

Grapevine canes are a less promising source of polyphenols to enrich grape must than grape pomace or grapevine leaves. Polyphenol extraction from canes provided just a slight increase in the concentration and number of compounds (Table 2). Caftaric, protocatechuic, and coumaric acid were the predominant phenolic acids provided by PA and LC canes, respectively. The flavonols myricetin-3-hexoside, quercetin-3-hexoside, and quercetin-3-glucuronide, which are present at low concentrations, were from cane origin in case of Lachryma Christi. However, these compounds were also detected in pomace and leaves (Table 1). The presence of these compounds was in agreement with previous reports identifying caftaric acid, epicatechin, and quercetin and malvidin derivatives as the main metabolites concerning grape stems [40,41]. Although the presence of stilbenes in grape canes has been reported previously [42–44], these compounds were not detected in this study. The absence of stilbenoids in grapevine extracts can be attributed to the use of polar aqueous extraction conditions and the lack of UV-C irradiation in grapevines, as the biosynthesis of resveratrol in grape leaves is strongly increased in response to UV-C irradiation [45], or the storage at −20 °C of grapevine canes after collection. E-resveratrol levels in fresh cut canes that have been kept frozen were very low, whereas post-pruning storage at room temperature induced E-resveratrol biosynthesis, giving a significant rise of E-resveratrol levels after several months [43].

3.2. Physicochemical Characterization

Physicochemical differences were observed between enriched grape juice, i.e., a mixture of pomace, leaf, and cane extracts, and base grape musts due to the extraction conditions and the addition of compounds from leaf, cane, and pomace (Table 3).

Table 3. Physicochemical properties and chemical composition of grape musts and enriched juice [1].

	PA Must	LC Must	Enriched Juice
pH	3.17 ± 0.01 b	2.88 ± 0.01 c	3.43 ± 0.02 a
Total acidity (g/kg)	2.46 ± 0.03 c	4.81 ± 0.05 a	2.73 ± 0.00 b
Soluble solids (Brix)	17.8 ± 0.0 a	15.9 ± 0.0 c	17.5 ± 0.1 b
Free SO_2 (mg/kg)	9.6 ± 0.0 b	25.6 ± 3.2 a	1.5 ± 0.0 c
Total SO_2 (mg/kg)	19.2 ± 0.0 b	41.6 ± 2.8 a	1.7 ± 0.3 c
Color intensity	1.49 ± 0.01 c	2.95 ± 0.03 b	4.64 ± 0.09 a
Hue—tint	1.37 ± 0.02 a	0.40 ± 0.01 c	0.92 ± 0.02 b
Total polyphenols (mg/kg)	763 ± 17 c	2015 ± 170 a	1559 ± 59 b
Total flavonoids (mg/kg)	711 ± 66 b	995 ± 351 ab	1326 ± 32 a
Monomeric anthocyanins (mg/kg)	0.70 ± 0.00 c	218 ± 8 a	61.0 ± 3.7 b
ABTS (mmol/kg)	9.59 ± 0.98 b	21.5 ± 4.6 b	77.2 ± 11.1 a

[1] Data are expressed as mean ± standard deviation (*n* = 3). Different letters in the same row indicates statistically significant difference (*p* < 0.05). PA: cultivar País, LC: cultivar Lachryma Christi.

Unripe grapes were used to prepare musts according to ripeness criteria of soluble solids, acidity, and pH for wine grapes [46]. Increased pH values and a relatively low total acidity found in enriched juice may be due to the loss of volatile compounds during thermomaceration. On the other hand, Lachryma Christi grapes result in an unusual juice that has been characterized by a relatively low sugar content, high acidity, low pH, and high color intensity. Moreover, these grapes are a better source of phenolics, in particular, monomeric anthocyanins and total polyphenols, when compared to País grapes (Table 3). Additionally, total antioxidant capacity evaluated by ABTS$^+$ scavenging showed a significant increase for enriched juice when compared to starting materials. These results show the potential of grape pomace, grapevine leaves and canes' residues as new resources of antioxidants to enrich fruit juices by using green extraction techniques. Relatively high free bisulfite concentrations of base grape musts are important to stabilize them by preventing enzymatic browning and the deterioration of aroma perception. Bisulfite is able to protect grape musts against enzymatic oxidation by polyphenol oxidase via the formation of 2-*S*-glutathione caftaric acid, which is a relatively stable grape reaction product [47]. This protective role may avoid the formation of *O*-quinones derived from hydroxycinnamic acids and catechins that are known to react easily with cysteinylated aroma precursors from grape juice and skins affecting aroma stability, as well as the formation of brown oligomers [48]. Since the solid fractions used in juice enrichment contain additional flavonoids, this results in a decline in protective bisulfite contents. Moreover, the relatively high process temperature and surface aeration by intense stirring may facilitate some loss of bisulfite as well. Analysis of color according to intensity and hue shows differences among samples. País had a more yellow hue, while the color of Lachryma Christi and fortified juice samples shifted towards reddish hue. Values of hue-tint and monomeric anthocyanin concentrations showed an inverse linear relationship (Pearson coefficient r = −0.976; $p < 0.01$). Anthocyanins and in particular malvidin-3-glucoside were the main pigments in grape juices with a positive correlation between malvidin-3-glucoside concentration and color stability [49]. The relative loss of monomeric anthocyanins after thermomaceration for enriched juice, in combination with an increase of color intensity, indicates co-pigmentation between anthocyanins and flavonoids extracted from skins, seeds, leaves, and canes. Co-pigmentation is an important phenomenon in the case of Teinturier grape cultivars [34]. Acylated anthocyanins are the main pigments of the co-pigmentation complex with cofactors, such as hydroxycinnamic acids (coumaric, caffeic, and ferulic acid), flavonol derivatives (myricetin, quercetin, and kaempferol), or flavone derivatives (vitexin and orientin) [34].

3.3. Phenolics as Antioxidant Agents

The antioxidant capacity of enriched juice in this study, as measured by the ability of antioxidants to scavenge ABTS$^+$ radicals, was found ~3.6- and ~8.0-fold higher than for base grape musts (Table 3) and exceeded the values reported for commercial white, red, and purple grape juices at least three-fold [9,50]. Values of antioxidant capacity did not show a linear relationship with monomeric anthocyanin contents (Pearson coefficient r = −0.080; $p = 0.837$) and total polyphenol contents (r = 0.317; $p = 0.407$), contrary to flavonoid contents (r = 0.743; $p = 0.022$). The increase of antioxidant capacity was stronger than the increment of monomeric anthocyanins and total polyphenols found after thermomaceration. This suggests that grape pomace, grapevine leaves and canes contain specific phenolics, whose molecular structures show a stronger antioxidant capacity than those that are contained in base grape musts, or that synergism occurs between them. Both the configuration and the number of hydrogen-donating hydroxyl groups are the main structural features influencing the antioxidant activity of polyphenols. For example, the high activity of catechin dimers among flavan-3-ols was attributed to their hydroxyl functional groups that are potent hydrogen donators [13]. Additionally, the ortho-dihydroxy structure on the B-ring, the 2-3-double bound conjugated with a 4-oxo function in the C-ring, and the free hydroxyl groups in position 3 in the C-ring and position 5 in the A-ring are important structural features for flavonoids [13]. This may explain the relatively high antioxidant activity of quercetin among flavonols, while the glycosylation of hydroxyl substituents on C3 will drop the antioxidant activity when compared to the aglycon [51]. Conjugated double bounds in combination with a planar molecular structure of quercetin allow for electron delocalization across the molecule, thus stabilizing the corresponding phenoxyl radicals [52]. In our study, both catechin dimers and quercetin derivatives increased in grapevine extracts, being the main ingredients of enriched grape juice. In addition, polyphenols are believed to scavenge free radicals by two major mechanisms: by reduction via electron transfer and by hydrogen atom transfer, which may occur in parallel [53]. The CUPRAC assay determines the antioxidant capacity of hydrophilic and lipophilic dietary polyphenols in vitro based on the single electron transfer principle. The ORAC assay evaluates the capacity of antioxidants to inhibit bleaching of a target molecule (probe) induced by peroxyl radicals according to the principle of hydrogen atom transfer. Therefore, it is important to run multiple antioxidant methods, rather than just the ABTS method to get a better estimate of antioxidant potency of phenolic-rich foods on human health.

Enriched grape juice exhibited ORAC-FL values of 19.6 ± 0.0 mmol/kg, being similar to those reported for red grape juices (14.6–25.0 mmol/kg) [53], while ORAC-PGR values (13.7 ± 0.1 mmol/kg) were between those that were reported for white and red wines [30]. When comparing both assays, the stoichiometry of reactions is more important than the reactivity of antioxidants for ORAC-FL, while the absence of induction times in the kinetic profiles of PGR consumption would imply that the ORAC-PGR index is more related to the reactivity of antioxidants than to stoichiometric factors. ORAC-FL is a measure of the amount of reactive polyphenols available, while ORAC-PGR are influenced by the quality of antioxidants present in the sample. The ORAC-PGR/ORAC-FL ratio would reflect the average quality of antioxidants present in a beverage. A high value of this ratio would imply that a high proportion of antioxidants that are contained in the sample are able to protect PGR against bleaching induced by peroxyl radicals. The ORAC-PGR/ORAC-FL ratio of 0.70 for enriched grape juice was similar to that of red, rosé, and white wines [54], but was 5 and 45 times higher than the values of tea and herbal infusions, respectively [55]. Thus, the reactivity of polyphenols that are present in enriched grape juice is at least as efficient as those present in other beverages that are rich in antioxidants.

According to the CUPRAC assay, the antioxidant capacity of enriched grape juice was 3.62 ± 0.20 mmol Trolox equivalents/kg sample. CUPRAC-measured antioxidant capacity was significantly lower than the ORAC values, indicating that enriched grape juice has a more potent radical scavenging capacity by hydrogen atom transfer than reducing capacity via electron transfer. Similar findings were reported after comparing ABTS- and FRAP-measured antioxidant capacities [9].

3.4. Sensorial Evaluation

The results of sensorial tests for LC extracts showed a significantly lower preference score of panelists for leaf extract than for pomace or cane extracts (Table 4).

Strong herbal, tea-like notes and astringency of leaf extracts were considered as negative attributes. These attributes were less notorious in case of PA extracts due to an increased sweetness, lower acidity, and greater aroma complexity that may suppress negative sensory perception. High amounts of flavan-3-ols found in LC leaf extracts can affect taste of this beverage. These compounds, in particular oligomeric tannins, are related to astringency and bitterness [56].

As evidenced by the preference ranking scores (Table 4), LC juice supplemented with pomace, leaves and canes at a ratio of 40:20:40 ($v/v/v$) was more attractive than other samples giving a complex aroma and a well-balanced sweet acid taste, combined with a slight astringency. In addition, half of País and half of Lachryma Christi juice enriched with pomace, leaves and canes at a ratio of 40:20:40 ($v/v/v$) should be mixed to yield a well-balanced taste of sweetness and acidity, together with slight notes of a vegetal-woody odor and a reddish hue with increased color intensity.

Table 4. Preference ranking score for blends of pomace, leaves and canes extracts [1].

First Level Target	Second Level Target	Score
País	Pomace	7.5 ± 1.2 a
	Leaf	4.0 ± 1.0 a
	Cane	3.5 ± 0.9 a
Lachryma Christi	Pomace	6.5 ± 1.1 a
	Leaf	1.0 ± 0.6 b
	Cane	7.5 ± 1.3 a
País	Pomace/leaf/cane ratio of 25/25/50	5.0 ± 0.5 a
	Pomace/leaf/cane ratio of 40/20/40	3.0 ± 0.5 a
	Pomace/leaf/cane ratio of 50/17/33	2.0 ± 0.4 a
Lachryma Christi	Pomace/leaf/cane ratio of 25/25/50	1.0 ± 0.3 b
	Pomace/leaf/cane ratio of 40/20/40	7.0 ± 0.5 a
	Pomace/leaf/cane ratio of 50/17/33	2.0 ± 0.4 b
Pomace/leaf/cane ratio of 40/20/40	País/Lachryma Christi ratio of 90/10	0.8 ± 0.2 b
	País/Lachryma Christi ratio of 50/50	5.4 ± 0.5 a
	País/Lachryma Christi ratio of 70/30	3.9 ± 0.5 ab

[1] Different letters in the same block indicates statistically significant difference ($p < 0.05$).

4. Conclusions

The polyphenol-rich grape juice, made by thermomaceration using grape pomace, grapevine leaves and canes, shows a promissory Trolox equivalent antioxidant capacity of 77.2 mmol/kg juice, according to the ABTS assay that may offer health benefits. Catechin dimers and quercetin derivatives with high antioxidant activity are main ingredients of enriched grape juice. Both grape pomace and grapevine leaves are of primary importance as additional polyphenol sources in the preparation of enriched grape juice. Pomace and leaf extracts of both País and Lachryma Christi cultivars evidenced an elevated concentration of phenolic compounds, which is largely attributed to their anthocyanin, flavonol, and flavan-3-ol content.

Acknowledgments: This work was supported by grants from the National Commission for Scientific and Technological Research (FONDEF VIU 120010 and CONICYT PFB.27). Viña Zamora provided grape samples.

Author Contributions: Tabita Aguilar conceived and designed the experiments; Cristina Loyola, Luis Bustamante and Carola Vergara performed the experiments; Tabita Aguilar, Dietrich von Baer, Ignacio Serra and Johannes de Bruijn analyzed the data; Claudia Mardones contributed reagents/materials/analysis tools; Johannes de Bruijn wrote the paper.

Conflicts of Interest: The authors declare no conflict of interest.

References

1. Renaud, S.C.; de Lorgeril, M. Wine, alcohol, platelets, and the French paradox for coronary heart disease. *Lancet* **1992**, *339*, 1523–15261. [CrossRef]
2. Folts, J.D. Potential health benefits from the flavonoids in grape products on vascular disease. In *Flavonoids in Cell Function*; Buslig, B., Manthey, J., Eds.; Kluwer Academic/Plenum Publishers: New York, NY, USA, 2002; pp. 95–111. ISBN 978-1-4757-5235-9.
3. Booyse, F.M.; Pan, W.; Grenett, H.E.; Parks, D.A.; Darley-Usmar, V.M.; Bradley, K.M.; Tabengwa, E.M. Mechanism by which alcohol and wine polyphenols affect coronary heart disease risk. *Ann. Epidemiol.* **2007**, *17*, S24–S31. [CrossRef] [PubMed]
4. Leifert, W.R.; Abeywardena, M.Y. Cardioprotective actions of grape polyphenols. *Nutr. Res.* **2008**, *28*, 729–737. [CrossRef] [PubMed]
5. Ruxton, C.H.; Gardner, E.J.; Walker, D. Can pure fruit and vegetable juices protect against cancer and cardiovascular disease too? A review of the evidence. *Int. J. Food Sci. Nutr.* **2006**, *57*, 249–272. [CrossRef] [PubMed]
6. Mullen, W.; Marks, S.C.; Crozier, A. Evaluation of phenolic compounds in commercial fruit juices and fruit drinks. *J. Agric. Food Chem.* **2007**, *55*, 3148–3157. [CrossRef] [PubMed]

7. Spanos, G.A.; Wrolstad, R.E. Phenolics of apple, pear, and white grape juices and their changes with processing and storage. A review. *J. Agric. Food Chem.* **1992**, *40*, 1478–1487. [CrossRef]
8. Pereira Natividade, M.M.; Corrêa, L.C.; Carvalho de Souza, S.V.; Pereira, G.E.; de Oliveira Lima, L.C. Simultaneous analysis of 25 phenolic compounds in grape juice for HPLC: Method validation and characterization of São Francisco Valley samples. *Microchem. J.* **2013**, *110*, 665–674. [CrossRef]
9. Moreno-Montoro, M.; Olalla-Herrera, M.; Gimenez-Martinez, R.; Navarro-Alarcon, M.; Rufián-Henares, J.A. Phenolic compounds and antioxidant activity of Spanish commercial grape juices. *J. Food Comp. Anal.* **2015**, *38*, 19–26. [CrossRef]
10. Krikorian, R.; Boespflug, E.L.; Fleck, D.E.; Stein, A.L.; Wightman, J.D.; Shidler, M.D.; Sadat-Hossieny, S. Concord grape juice supplementation and neurocognitive function in human aging. *J. Agric. Food Chem.* **2012**, *60*, 5736–5742. [CrossRef] [PubMed]
11. Toaldo, I.M.; Cruz, F.A.; da Silva, E.L.; Bordignon-Luiz, M.T. Acute consumption of organic and conventional tropical grape juices (*V. labrusca* L.) increases antioxidants in plasma and erythrocytes, but not glucose and uric acid levels in healthy individuals. *Nutr. Res.* **2016**, *36*, 808–817. [CrossRef] [PubMed]
12. Castilla, P.; Echarri, R.; Dávalos, A.; Cerrato, F.; Ortega, H.; Teruel, J.L.; Fernández-Lucas, M.; Gómez-Coronado, D.; Ortuño, J.; Lasunción, M.A. Concentrated red grape juice exerts antioxidant, hypolipidemic, and antiinflammatory effects in both hemodialysis patients and healthy subjects. *Am. J. Clin. Nutr.* **2006**, *84*, 252–262. [PubMed]
13. Soobratee, M.A.; Neergheena, V.S.; Luximon-Ramma, A.; Aruoma, O.I.; Bahorun, T. Phenolics as potential antioxidant therapeutic agents: Mechanism and actions. *Mut. Res.* **2005**, *579*, 200–213. [CrossRef] [PubMed]
14. Freedman, J.E.; Parker, C.; Li, L.; Perlman, J.A.; Frei, B.; Ivanov, V.; Deak, L.R.; Lafrati, M.D.; Folts, J.D. Select flavonoids and whole juice from purple grapes inhibit platelet function and enhance nitric oxide release. *Circulation* **2001**, *103*, 2792–2798. [CrossRef] [PubMed]
15. Corder, R.; Mullen, W.; Khan, N.Q.; Marks, S.C.; Wood, E.G.; Carrier, M.J.; Crozier, A. Red wine procyanidins and vascular health. *Nature* **2006**, *444*, 566. [CrossRef] [PubMed]
16. Faria, A.; Calhau, C.; de Freitas, V.; Mateus, N. Procyanidins as antioxidants and tumor cell growth modulators. *J. Agric. Food Chem.* **2006**, *54*, 2392–2397. [CrossRef] [PubMed]
17. Yi, W.; Fischer, J.; Akoh, C.C. Study of anticancer activities of muscadine grape phenolics in vitro. *J. Agric. Food Chem.* **2005**, *53*, 8804–8812. [CrossRef] [PubMed]
18. Lu, R.; Serreno, G. Resveratrol, a natural product derived from grape, exhibits antiestrogenic activity and inhibits the growth of human breast cancer cells. *J. Cell. Physiol.* **1999**, *179*, 297–304. [CrossRef]
19. Nishiumi, S.; Mukai, R.; Ichiyanagi, T.; Ashida, H. Suppression of lipopolysaccharide and galactosamine-induced hepatic inflammation by red grape pomace. *J. Agric. Food Chem.* **2012**, *60*, 9315–9320. [CrossRef] [PubMed]
20. Zacharof, M.P. Winery waste as feedstock for bioconversions: Applying the biorefinery concept. *Waste Biomass Valorization* **2017**, *8*, 1011–1025. [CrossRef]
21. Teixeira, A.; Baenas, N.; Dominguez-Perles, R.; Barros, A.; Rosa, E.; Moreno, D.A.; García-Viguera, C. Natural bioactive compounds from winery-by-products as health promotors: A review. *Int. J. Mol. Sci.* **2014**, *15*, 15638–15678. [CrossRef] [PubMed]
22. Martínez, R.; Vaderrama, N.; Moreno, J.; de Bruijn, J. Aroma characterization of grape juice enriched with grapevine by-products using thermomaceration. *Chil. J. Agric. Res.* **2017**, *77*, 234–242. [CrossRef]
23. Aguilar, T.; Loyola, C.; de Bruijn, J.; Bustamante, L.; Vergara, C.; von Baer, D.; Mardones, C.; Serra, I. Effect of thermomaceration and enzymatic maceration on phenolic compounds of grape must enriched by grape pomace, vine leaves and canes. *Eur. Food Res. Technol.* **2016**, *242*, 1149–1158. [CrossRef]
24. International Organisation of Vine and Wine (OIV). *Compendium of International Methods of Wine and Must Analysis*; International Organisation of Vine and Wine: Paris, France, 2016; ISBN 979-10-91799-47-8.
25. Singleton, V.L.; Orthofer, R.; Lamuela-Raventós, R.M. Analysis of total phenols and other oxidation substrates and antioxidants by means of Folin-Ciocalteu reagent. *Methods Enzymol.* **1999**, *299*, 152–178. [CrossRef]
26. Lee, J.; Durst, R.W.; Wrolstad, R.E. Determination of total monomeric anthocyanin pigment content of fruit juices, beverages, natural colorants, and wines by the pH differential method: Collaborative study. *J. AOAC Int.* **2005**, *88*, 1269–1278. [PubMed]
27. Zhishen, J.; Mengcheng, T.; Jianming, W. The determination of flavonoid contents in mulberry and their scavenging effects on superoxide radicals. *Food Chem.* **1999**, *64*, 555–559. [CrossRef]

28. Ruiz, A.; Bustamante, L.; Vergara, C.; von Baer, D.; Hermosín-Gutiérrez, I.; Obando, L.; Mardones, C. Hydroxycinnamic acids and flavonols in native edible berries of South Patagonia. *Food Chem.* **2015**, 84–90. [CrossRef] [PubMed]

29. Wu, X.; Beecher, G.R.; Holden, J.M.; Haytowitz, D.B.; Gebhardt, S.E.; Prior, R.L. Lipophilic and hydrophilic antioxidant capacities of common foods in the United States. *J. Agric. Food Chem.* **2004**, *52*, 4026–4037. [CrossRef] [PubMed]

30. López-Alarcón, C.; Lissi, E. A novel and simple ORAC methodology based on the interaction of Pyrogallol Red with peroxyl radicals. *Free Radic. Res.* **2006**, *40*, 979–985. [CrossRef] [PubMed]

31. Muñoz, S.; Mestres, M.; Busto, O.; Guasch, J. Determination of some flavan-3-ols and anthocyanins in red grape seed and skin extracts by HPLC-DAD: Validation study and response comparison of different standards. *Anal. Chim. Acta* **2008**, *628*, 104–110. [CrossRef]

32. Nogales-Bueno, J.; Baca-Bocanegra, B.; Jara-Palacios, M.J.; Hernández-Hierro, J.M.; Heredia, F.J. Evaluation of the influence of White grape seed extracts as copigment sources on the anthocyanin extraction from grape skins previously classified by near infrared hyperspectral tools. *Food Chem.* **2017**, *221*, 1685–1690. [CrossRef] [PubMed]

33. Ghafoor, K.; Choi, Y.H.; Jeon, J.Y.; Jo, I.H. Optimization of ultrasound-assisted extraction of phenolic compounds, antioxidants, and anthocyanins from grape (*Vitis vinifera*) seeds. *J. Agric. Food Chem.* **2009**, *57*, 4988–4994. [CrossRef] [PubMed]

34. Boulton, R. The copigmentation of anthocyanins and its role in the color of red wine: A critical review. *Am. J. Enol. Vitic.* **2001**, *52*, 67–87.

35. Lutz, M.; Jorquera, K.; Cancino, B.; Ruby, R.; Henriquez, C. Phenolics and antioxidant capacity of table grape (*Vitis vinifera* L.) cultivars grown in Chile. *J. Food Sci.* **2011**, *76*, C1088–C1093. [CrossRef] [PubMed]

36. Davidov-Pardo, G.; Arozarena, I.; Marín-Arroyo, M.R. Stability of polyphenolic extracts from grape seeds after thermal treatments. *Eur. Food Res. Technol.* **2011**, *232*, 211–220. [CrossRef]

37. Capanoglu, E.; de Vos, R.C.H.; Hall, R.D.; Boyacioglu, D.; Beekwilder, J. Changes in polyphenol content during production of grape juice concentrate. *Food Chem.* **2013**, *139*, 521–526. [CrossRef] [PubMed]

38. Doshi, P.; Adsule, P.; Banerjee, K. Phenolic composition and antioxidant activity in grapevine parts and berries (*Vitis vinifera* L.) cv. Kismish Chornyi (Sharad Seedless) during maturation. *Int. J. Food Sci. Technol.* **2006**, *41*, 1–9. [CrossRef]

39. Liang, Z.; Owens, C.L.; Zhong, G.Y.; Cheng, L. Polyphenolic profiles detected in the ripe berries of *Vitis vinifera* germplasm. *Food Chem.* **2011**, *129*, 940–950. [CrossRef] [PubMed]

40. Barros, A.; Gironés-Vilaplana, A.; Teixeira, A.; Collado-González, J.; Moreno, D.A.; Gil-Izquierdo, A.; Rosa, E.; Domínguez-Perles, R. Evaluation of grapes (*Vitis vinifera* L.) stems from Portuguese varieties as a resource of (poly)phenolic compounds: A comparative study. *Food Res. Int.* **2014**, *65*, 375–384. [CrossRef]

41. Eftekhari, M.; Yadollahi, A.; Ford, C.M.; Shojaeiyan, A.; Ayyari, M.; Hokmabadi, H. Chemodiversity evaluation of grape (*Vitis vinifera*) vegetative parts during summer and early fall. *Ind. Crops Prod.* **2017**, *108*, 267–277. [CrossRef]

42. Vergara, C.; von Baer, D.; Mardones, C.; Wilkens, A.; Wernekinck, K.; Damm, A.; Macke, S.; Gorena, T.; Winterhalter, P. Stilbene levels in grape cane of different cultivars in Southern Chile: Determination by HPLC-DAD-MS/MS method. *J. Agric. Food Chem.* **2012**, *60*, 929–933. [CrossRef] [PubMed]

43. Gorena, T.; Saez, V.; Mardones, C.; Vergara, C.; Winterhalter, P.; von Baer, D. Influence of post-pruning storage on stilbenoid levels in *Vitis vinifera* L. canes. *Food Chem.* **2014**, *155*, 256–263. [CrossRef] [PubMed]

44. Balík, J.; Kyseláková, M.; Vrchotová, N.; Tříska, J.; Kumšta, M.; Veverka, J.; Híc, P.; Totušek, J.; Lefnerová, D. Relations between polyphenols content and antioxidant activity in vine grapes and leaves. *Czech J. Food Sci.* **2008**, *26*, S25–S32.

45. Xi, H.F.; Ma, L.; Wang, L.N.; Li, S.H.; Wang, L.J. Differential response of the biosynthesis of resveratrols and flavonoids to UV-C irradiation in grape leaves. *N. Z. J. Crop Hort.* **2015**, *43*, 163–172. [CrossRef]

46. Coombe, B.G.; Dundon, R.J.; Short, A.W.S. Indices of sugar-acidity as ripeness criteria for winegrapes. *J. Sci. Food Agric.* **1980**, *31*, 495–502. [CrossRef]

47. Cheynier, V.F.; Trousdale, E.K.; Singleton, V.L.; Salgues, M.J.; Wylde, R. Characterization of 2-S-glutathionyl caftaric acid and its hydrolysis in relation to grape wines. *J. Agric. Food Chem.* **1986**, *34*, 217–221. [CrossRef]

48. Maggu, M.; Winz, R.; Kilmartin, P.A.; Trought, M.C.T.; Nicolau, L. Effect of skin contact and pressure on the composition of Sauvignon Blanc must. *J. Agric. Food Chem.* **2007**, *55*, 10281–10288. [CrossRef] [PubMed]

49. Lambri, M.; Torchio, F.; Colangelo, D.; Río Segade, S.; Giacosa, S.; De Faveri, D.M.; Gerbi, V.; Rolle, L. Influence of different berry thermal treatment conditions, grape anthocyanin profile, and skin hardness on the extraction of anthocyanin compounds in the colored grape juice production. *Food Res. Int.* **2015**, *77*, 584–590. [CrossRef]

50. Seeram, N.P.; Aviram, M.; Zhang, Y.; Henning, S.M.; Feng, L.; Dreher, M.; Heber, D. Comparison of antioxidant potency of commonly consumed polyphenol-rich beverages in the United States. *J. Agric. Food Chem.* **2008**, *56*, 1415–1422. [CrossRef] [PubMed]

51. Villaño, D.; Fernández-Pachón, M.S.; Troncoso, A.M.; García-Parilla, M.C. Comparison of antioxidant activity of wine phenolic compounds and metabolites in vitro. *Anal. Chim. Acta.* **2005**, *538*, 391–398. [CrossRef]

52. Van Acker, S.A.B.E.; de Groot, M.J.; van den Berg, D.J.; Tromp, M.N.J.L.; Donné-Op den Kelder, G.; van der Vijgh, W.J.F.; Bast, A. A quantum chemical explanation of the antioxidant activity of flavonoids. *Chem. Res. Toxicol.* **1996**, *9*, 1305–1312. [CrossRef] [PubMed]

53. Dávalos, A.; Bartolomé, B.; Gómez-Cordovés, C. Antioxidant properties of commercial grape juices and vinegars. *Food Chem.* **2005**, *93*, 325–330. [CrossRef]

54. López-Alarcón, C.; Ortíz, R.; Benavides, J.; Mura, E.; Lissi, E. Use of the ORAC-Pyrogallol Red/ORAC-Fluorescein ratio to assess the quality of antioxidants in Chilean wines. *J. Chil. Chem. Soc.* **2011**, *56*, 764–767. [CrossRef]

55. Alarcón, E.; Campos, A.M.; Edwards, A.M.; Lissi, E.; López-Alarcón, C. Antioxidant capacity of herbal infusions and tea extracts: A comparison of ORAC-fluorescein and ORAC-pyrogallol red methodologies. *Food Chem.* **2008**, *107*, 1114–1119. [CrossRef]

56. Toaldo, I.M.; Fogolari, O.; Cadore Pimentel, G.; Santos de Gois, J.; Borges, D.L.G.; Caliari, V.; Bordignon-Luiz, M.T. Effect of grape seeds on the polyphenol bioactive content and elemental composition by ICP-MS of grape juices from *Vitis labrusca* L. *LWT Food Sci. Technol.* **2013**, *53*, 1–8. [CrossRef]

beverages

MDPI

Article

Optimization of the Juice Extraction Process and Investigation on Must Fermentation of Overripe Giant Horn Plantains

C. W. Makebe [1], Z. S. C. Desobgo [2,*] and E. J. Nso [1]

[1] Department of Process Engineering, National School of Agro-Industrial Sciences (ENSAI), University of Ngaoundere, P.O. Box 455 Ngaoundere, Cameroon; cally_maya@yahoo.com (C.W.M.); nso_emmanuel@yahoo.fr (E.J.N.)

[2] Department of Food Processing and Quality Control, University Institute of Technology (UIT), The University of Ngaoundere, P.O. Box 455 Ngaoundere, Cameroon

Academic Editor: António Manuel Jordão
Received: 6 March 2017; Accepted: 6 April 2017; Published: 13 April 2017

Abstract: The study was initiated to optimize the enzymabtic extraction process of plantain pulp using response surface methodology. Weight loss of plantain decreased until it became stable at an over-ripe stage. The significant regression model describing the changes of extraction yield and Brix with respect to hydrolysis parameters was established. Temperature contributed to reducing the yield from 53.52% down to 49.43%, and the dilution factor increased the yield from 53.52% to 92.97%. On the contrary, the dilution factor significantly reduced Brix from 21.74 °Bx down to 0.15 °Bx, while the enzyme concentration increased Brix from 21.73 °Bx to 26.16 °Bx. The optimum conditions for juice extraction from plantain pulp were: temperature: 25 °C; enzyme concentration: 5%; dilution ratio: 1.10; and extraction time: 24 h. The implementation of these conditions led to (resulted in obtaining) obtaining a must yield of more than 70% and Brix between 10 °Bx and 15 °Bx. The total polyphenols and flavonoids were 7.70 ± 0.99 mg GAE /100 g and 0.4 ± 0.01 µg rutin/g for must and 17.01 ± 0.34 mg GAE/100 g and 4 ± 0.12 µg rutin/g and 7.70 ± 0.99 for wine, indicated the presence of antioxidant activity in the produced wine. On the other hand, the total soluble solids were between 16.06 ± 0.58 °Bx and 1.5 ± 0.10 °Bx, which permitted obtaining a wine with low alcohol content.

Keywords: plantains; losses; extraction; response surface methodology; wine

1. Introduction

Although banana/plantain is a staple for many African countries, it has until recently not been successfully exploited in food processing and assimilated industries due to the lack of adequate post-harvest preservation technologies [1]. The high level of post-harvest losses is one of the food security challenges facing African countries. In Sub-Saharan Africa, the annual value of these losses is estimated at about USD 48 billion [2]. Cameroon produces about 3,882,741 million tons of plantain annually [3]. Approximately half of this is consumed locally, while less than two fifths is exported. The remaining quantities rot in the fields because of rapid ripening of the fruits, poor handling, inadequate storage and transportation means and poor knowledge of food processing options [1]. It is estimated that more than 30% of the banana [4] and 35% of plantain [5] production are lost after harvest in developing countries. In Cameroon, studies have estimated that 30% of post-harvest losses are incurred during whole sale and about 70% during retailing [4].

Banana plants are important monocotyledonous perennial crops in the tropical and subtropical world regions [6]. They include dessert banana, plantain and cooking bananas. Plantain (*Musa paradisiaca*) is one of the most important crops of the tropical plants. It belongs to the family Musaceae and the genus *Musa*. Lately, wine production has been honed with different natural products, for example apple, pear and strawberry, fruits, plum, pineapple, banana and oranges [7–9]. Wines are energizing drinks that have been viewed as a characteristic solution for man's disease from early days and are said to help recuperation amid the gaining strength period [10,11]. Studies to upgrade banana juice extraction were completed by utilizing diverse levels of pectinase enzymes and distinctive hatching periods at 28 ± 2 °C [12]. The juice recuperation from over-ripe bananas was higher (67.6%) than that from ordinary natural fruits (60.2%). A great quality wine was obtained from over-ripe banana fruits [12,13].

There is an extensive misuse of bananas, particularly those that do not meet the quality standards for export. Because of the way that the banana/plantain fruit is organically dynamic and completes transpiration, ripening and other biochemical exercises even after harvest, the losses are high [5]. This phenomenon has shifted to break down the nature of leafy foods, making them unmarketable. Wastage of the plantain fruit because of poor post-harvest handling or over-ripening remains a noteworthy issue today. The change of over-ripe plantain into other important items has been the main impetus behind this study. Over the span of this study, we chose to concentrate on the production of wine from giant horn over-ripe plantains. This will be done using response surface methodology and by applying industrial enzymes to quantitatively extract plantain juice, determining the parameters for optimum yield and analyzing some physicochemical parameters before and after fermentation.

2. Material and Methods

2.1. Material

2.1.1. Biological Material

Giant horn plantain (Figure 1) was obtained from a farmer in Mamfe in the South West Region, Cameroon. Enzymes' characteristics utilized were as mentioned in Table 1. The white wine yeast Lalvin ICV D-47 *Saccharomyces cerevisiae* was issued by The Home Brew Shop (Unit 3, Hawley Lane Business Park, United Kingdom).

Figure 1. Giant horn plantains ((**A**) unripe; (**B**) over-ripe).

Table 1. Enzyme properties.

Enzymes (Origins, Suppliers)	Enzymes Percentage in the Mixture	pH		Temperature	
		Tolerance	Optimal	Tolerance	Optimal
Amyloglucosidase (AMG 300 L, Novozymes)	43%	4.0–6.0	4.5	30–65 °C	60 °C
Alpha-amylase (BAN 480 L, Novozymes)	29%	5.0–6.0	5.4–5.8	37–75 °C	70 °C
Pectinase (*Aspergillus aculeatus*, Sigma Aldrich)	14%	3.5–6.0	4.5	30–55 °C	50 °C
Bioglucanase TX (*Trichoderma reesei*, Kerry Bioscience)	14%	4.5–6.5	4.5–6.5	/	60 °C

2.1.2. Chemicals

The following reagents were obtained from Fluka Chemika, Switzerland: 3,5-dinitrosalicylic acid (DNS), ethanol, fructose, D-Glucose, NaOH and maltose; while ninhydrin, sodium carbonate (Na_2CO_3), sodium potassium tartrate and sodium hydroxide were obtained from Chemphora Chemical, Netherlands.

2.2. Methods

2.2.1. Sample Ripening

Unripe mature plantains were over-ripened for 11 days at 25 ± 2 °C utilizing a Heraeus-type incubator (D-63450 Hanau, Germany) without the use of ethylene until they passed Stage 7 (over-ripe). After that, samples were used immediately for wine production.

2.2.2. General Process Overview

The methods involved in wine production from over-ripe giant horn plantain are given in Figure 2.

The over-ripe plantains coming from the ripening process were washed three times with tap water and peeled. After the peeling process, they were sized and crushed to get the pulp paste. Then, the blanching process was executed at 80 °C for 5 min. Pulp paste was mashed using Doehlert experimental design conditions when the enzyme mixture, sodium bicarbonate and water were added. At the end of the mashing process, the mash was centrifuged ($6500\times g$, 15 min) to remove pellets, and the must was then pasteurized (90 °C, 3 min). After cooling, the must was fermented at 20 °C for 7 days using yeast (4 g/L) and aged for 90 days. At the end of the aging process, the wine was centrifuged ($5000\times g$, 5 min) to remove pellets (trub and yeast) and bottled. The bottles containing wine were then pasteurized at 65 °C for 30 min.

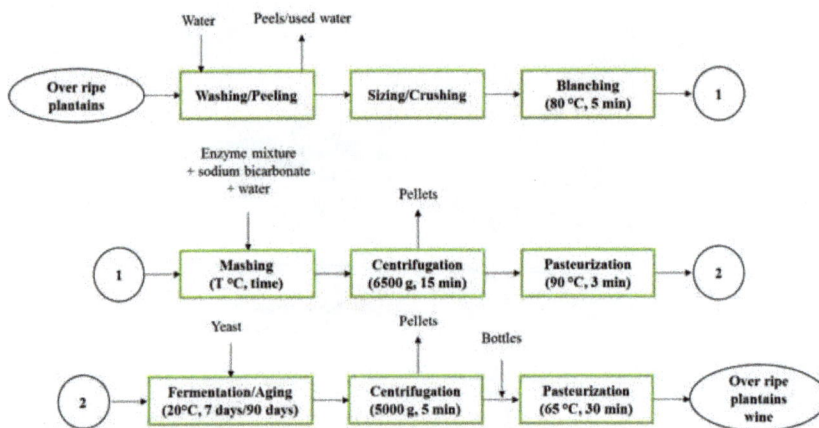

Figure 2. Process diagram for the ripe plantain wine production.

2.2.3. Experimental Design for the Extraction Process

Doehlert's experimental design [14] was used to model four factors: extraction temperature (°C), enzyme concentration (%), dilution factor and extraction time (h). The experimental domains (Table 2) over which factors were studied were based on the literature. These independent variables were studied using the response surface methodology (RSM) at three different levels (-1, 0, +1) to give a total of 25 experiments. The conversion of coded values into real values was done according to the literature [15].

Table 2. Experimental level of factors for ripe plantain fermentation.

Factors (Variables)	Level
Temperature (°C)	25–75
Enzyme mix concentration (%)	0.1–5
Dilution factor	0–4
Time (h)	0.5–24

2.2.4. Modelling and Optimizing

The model chosen was an empirical quadratic model, which determined the magnitude of the contributions of each factor in the linear, quadratic and interaction forms. The model used to determine the responses is given below:

$$Y_i = \beta_0 + \sum \beta_i x_i + \sum \beta_{ii} x_i^2 + \sum \beta_{ij} x_i x_j + \varepsilon \tag{1}$$

where Y_i is the response, β_0 a constant, $\beta_1, \beta_2, \beta_3$ the linear coefficients, $\beta_{11}, \beta_{22}, \beta_{33}$ the quadratic coefficients, $\beta_{12}, \beta_{23}, \beta_{13}$ the coefficients of interaction, and $x_1, x_2, x_3, x_1 x_2, x_1 x_3, x_2 x_3$ and x_1^2, x_2^2, x_3^2, the linear, interaction and quadratic forms of the experimental variables respectively. ε is the error. The determination of the coefficients of the models was done by a multilinear regression in which we had: N the number of experiments in the design matrix, p the number of coefficients in the chosen model, y the vector of experimental responses y_i, $y = (y_1, y_2, ... y_N)$; η the vector of theoretical responses η_i, $\eta = (\eta_1, \eta_2, ..., \eta_N)$; X the matrix $N \times p$, matrix of the model.

$$X = \begin{pmatrix} 1 & x_{11} & x_{12} & \cdot & \cdot & x_{1,p-1} \\ 1 & x_{21} & x_{22} & \cdot & \cdot & x_{2,p-1} \\ & \cdot & & \cdot & & \cdot \\ 1 & \cdot & \cdot & & \cdot & \cdot \\ & \cdot & & & \cdot & \cdot \\ 1 & x_{N1} & x_{N2} & \cdot & \cdot & x_{N,p-1} \end{pmatrix} \tag{2}$$

β is the column vector of the parameters to be estimated; β_i ... ; $\beta = (\beta_1, \beta_2, \beta_3, ..., \beta_{p-1})$; $\hat{\beta}$ is the column vector of estimators $\hat{\beta}_i$, ε is the column vector of experimental error e_i, $\varepsilon = (e_1, e_2, ..., e_N)$.

Given the following classical matrix:

$$\eta = X\beta \tag{3}$$

$$y = X\beta + \varepsilon \tag{4}$$

$$\begin{pmatrix} y_1 \\ y_2 \\ \cdot \\ \cdot \\ \cdot \\ y_N \end{pmatrix} = \begin{pmatrix} 1 & x_{11} & x_{12} & \cdot & \cdot & x_{1,p-1} \\ 1 & x_{21} & x_{22} & \cdot & \cdot & x_{2,p-1} \\ & \cdot & & \cdot & & \cdot \\ 1 & \cdot & \cdot & & \cdot & \cdot \\ & \cdot & & & \cdot & \cdot \\ 1 & x_{N1} & x_{N2} & \cdot & \cdot & x_{N,p-1} \end{pmatrix} \begin{pmatrix} \beta_0 \\ \beta_1 \\ \cdot \\ \cdot \\ \cdot \\ \beta_N \end{pmatrix} + \begin{pmatrix} e_1 \\ e_2 \\ \cdot \\ \cdot \\ \cdot \\ e_N \end{pmatrix} \tag{5}$$

when $N > p$, the least squares method is used to estimate the coefficients of the polynomial.

$$\hat{\beta} = (X'X)^{-1} X'Y \tag{6}$$

$\hat{\beta}$ is an estimator of β

To this effect, Minitab software was used to determine the coefficients of the models.

In order to write equations that allow for the prediction of responses in the domain defined for the studies, it was important to validate the models. To that effect, the observed values and the values predicted by the model were compared. Further statistical tools were used in addition to the determination coefficient. Absolute average deviation (AAD) was calculated as follows:

$$AAD = \frac{\sum_{i=1}^{p}\left(\frac{|Y_{i\exp}-Y_{itheo}|}{Y_{i\exp}}\right)}{N} \tag{7}$$

$Y_{i\exp}$ is the experimental response and Y_{itheo} the response calculated from model i; N is number of experiments.

The bias factor (B_f) and exactitude factor (A_f) were also expressed as follows:

$$B_f = 10^B \tag{8}$$

$$A_f = 10^A \tag{9}$$

with:

$$B = \frac{1}{N}\sum_{i=1}^{N}\log\left(\frac{Y_{itheo}}{Y_{i\exp}}\right) \tag{10}$$

$$A = \frac{1}{N}\sum_{i=1}^{N}\left|\log\left(\frac{Y_{itheo}}{Y_{i\exp}}\right)\right| \tag{11}$$

Table 3 gives the acceptable values of the different indicators of valid models. The graphs and contour plots were done using OriginPro 2016 b9.3.226 (OriginLab Corporation, www.originlab.com).

Optimization was done using the software Minitab® 17 (Build 17.3.1, Minitab, Inc.). The conditions were fixed to maximize the extraction yield and Brix. After that, a compromise was made. The use of Sigmaplot (Version 12.5, Systat Software, Inc.) allowed drawing the contour plots and superimpose the graphs in order to determine the interest zone.

Table 3. Standard and acceptable values of the indicators of the validation of the models.

Indicators of Validation	Standard Values	Acceptable Values	References
Adjusted R^2	1	≥ 0.8	[16]
AAD	0	[0–0.3]	[17]
Bias factor	1	[0.75–1.25]	[18]
Exactitude factor	1	[0.75–1.25]	[18]

The analysis of variance (Table 4) was used to identify the influence of each factor and also the significance of their effects. This was achieved by comparing the average square of each effect to the experimental error. The significance of each factor was determined by the Fischer test. The significance of the Fischer (F) 'ratio' indicates the values necessary to reject the null hypothesis at the 0.05 probability level.

Table 4. Analysis of variance (ANOVA).

Source of Variation	DF	SS	Mean Square	Fisher (F-Test)
Regression (between)	$p-1$	SSR	$\frac{SSR}{p-1}$	$\frac{\frac{SSR}{p-1}}{\frac{SSE}{N-p}}$
Residue (within)	$N-p$	SSE	$\frac{SSE}{N-p}$	
Validity	$(N-p)-(n_0-1)$	SSE_2	$\frac{SSE_2}{(N-p)-(n_0-1)}$	$\frac{\frac{SSE_2}{(N-p)-(n_0-1)}}{\frac{SSE_1}{n_0-1}}$
Error	(n_0-1)	SSE_1	$\frac{SSE_1}{n_0-1}$	
Total	$N-1$	TSS		

2.2.5. Juice Extraction

Over ripe plantains were washed 3 times, hand peeled and sized. A Panasonic blender (MX-GM1011) was used to crush the plantains, then they were blanched in water at 80 °C for 5 min in a water bath (Memmert). The four enzymes, amyloglucosidase, alpha amylase (BAN 480 L), pectinase and bioglucanase TX, were then introduced into biological reactors (beakers) containing 20 g of the pulp with its respective dilutions, which was then incubated at a temperature and time specified by the Doehlert experimental design in a water bath. After incubation, the pulp was then centrifuged at 6500× *g* for 15 min using a centrifuge (Heraeus-Kendro Lab products, model: Biofuge primo R, type: D-37520, Hanau, Germany). The supernatant was collected and pasteurized at 90 °C for 1 min in the water bath. Each experiment was conducted in triplicate.

2.2.6. Yeast Preparation and Must Fermentation

The white wine yeast Lalvin ICV D-47 *Saccharomyces cerevisiae* obtained from The Home Brew Shop (Unit 3, Hawley Lane Business Park, United Kingdom) was used for the fermentation of the giant horn plantain. Yeast, which was kept at 2–4 °C, was rehydrated at 30 °C in tap water (10 mL of water/g of yeast) for about 10 min and stirred for 5 s. After rehydration, yeast was introduced to the giant horn plantain juice extracted utilizing optimal conditions, and the mixture was mixed to ensure must aeration and yeast distribution. The principal fermentation was then done at 20 °C during 7 days and aging for 90 days in an incubator (SHP Biochemical Incubator 250 L, Shanghai, Guangzhou, China).

2.2.7. Determination of Extraction Yield

The percentage of juice yield (% w/w) was calculated as the difference between the initial mass and the weight of the pellet after centrifugation divided by the initial mass as follows:

$$Yield(\%) = \frac{mass\ of\ pulp\ -\ mass\ of\ pellet}{mass\ of\ pulp} \times 100 \tag{12}$$

2.2.8. Determination of Total Soluble Solids [19]

Before reading the value of total soluble solids, the Hanna HI 96801 refractometer (Hanna instruments Inc., Woonsocket, RI, USA) was calibrated using distilled water. An equal number of drops from the prepared plantain juice was placed onto the refractometer prism plate. The reading on the prism scale is generated numerically. After each test, the prism plate was cleaned with (distilled) water and wiped dry with a soft tissue.

2.2.9. Determination of pH

The initial pH of each plantain juice was determined using a pH meter (HANNA® Calibration Check™ pH meter, HI 223 Type, Johannesburg, South Africa). Twenty milliliters of each freshly-prepared plantain juice was placed in a glass beaker on a thermostatically-controlled electric hotplate (mark: GMARK) at 25 °C. Before reading its pH, each sample was agitated (using a magnetic stirrer) for 30 s until a stable reading was assessed. Each test was made in triplicate. Between readings, the electrode was rinsed with distilled water for the accuracy of the measurement.

2.2.10. Determination of Turbidity

To the turbidimeter cell, the plantain juice was added up to the horizontal mark. After closing the cell, it was wiped using a tissue and placed in the turbidimeter (Hach, Model 2100 N, Hach Company, Loveland, CO, USA), and the sample was covered. The turbidity value was then assessed when the reading was stable.

2.2.11. Determination of Total Phenolic Compounds: Folin–Ciocalteu Method [20]

The volume of 0.02 mL of plantain juice before and after fermentation was mixed with 0.1 mL of Folin–Ciocalteu reagent previously diluted with distilled water. A volume of 0.3 mL of 20% sodium carbonate solution was added to the mixture, shaken thoroughly and diluted to 2 mL by adding distilled water. The mixture was allowed to stand for 120 min, and the blue color formed was measured at 760 nm with a spectrophotometer (JENWAY, Model: 7310, Serial No. 39756, JENWAY Limited, Staffordshire, UK). Gallic acid was invoked as a standard for the calibration curve. The concentrations of gallic acid in the solution, utilized for obtaining the calibration curve, were 0, 1, 2, 3, 4, 5, 6, 7, 8, 9 and 10 mg/L (R^2 = 0.965). The total polyphenol content was expressed as mg of gallic acid equivalent (GAE) per 100 g of fruit juice. All measurements were carried out in triplicate.

2.2.12. Determination of Titratable Acidity [21]

Titratable acidity was established according to the standardized method, with 0.1 N sodium hydroxide (NaOH) in the presence of the phenolphthalein indicator. Ten milliliters of sample were pipetted into a conical flask and 0.1 mL of phenolphthalein (0.05%) added. Titration is halted when the initial color changes to pink and persisting for at least 30 s. The burette reading is noted. The titratable acidity (TA) is expressed in g/L tartaric acid:

$$TA = \frac{75 \times C \times V}{T} \qquad (13)$$

where: V is the volume (mL) of the sodium hydroxide noted at the end point; C is the concentration of the base; T is the volume of titrate.

2.2.13. Determination of Color by the Spectrophotometric Method [22]

The wavelength of the spectrophotometer (JENWAY, Model: 7310, Serial No. 39756, JENWAY Limited, Staffordshire, UK) was set at 430 nm. The cell was filled with water and the absorbance set to read as 0.00. The cell was then rinsed and filled with the sample (must and wine) and the absorbance read. For samples with a turbidity above 1 EBC, a filtration was done before using a 0.45-μm membrane filter (Whatmann, GE Healthcare, Chicago, IL, USA). The results were then expressed as follows:

$$Color\ (EBC) = A \times f \times 25 \qquad (14)$$

where: A is absorbance at 430 nm; f is the dilution factor; EBC is the European Brewery Convention units.

2.2.14. Dry Matter and Water (Moisture) Contents

Crushed plantain pulp (5 g) was weighed using a balance (SCIENTECH, ZSP250 MG Balance, Scientech Inc., Boulder, CO, USA) and placed in a drying dish. The dish was then placed in an oven (Heraeus, model: Kendro laboratory products, D-63450, GmbH, Hanau, Germany) previously set at 105 °C and left there till it attained constant weight. It was removed from the oven at the end of 24 h. After cooling in a desiccator, the dish was then weighed again. The dry matter content represents the difference in mass before and after drying in the oven. The dry matter (DM) content in 100 g of fresh sample was calculated using the following formula:

$$DM(\%) = \frac{M_2 - M_0}{M_1} \times 100 \qquad (15)$$

where: M_0 is mass (in g) of empty drying dish; M_1 is mass (in g) of sample before drying; M_2 is mass (in g) of drying dish + sample after drying.

The water or moisture content (%W) was calculated using the following expression:

$$W(\%) = 100 - DM \tag{16}$$

2.2.15. Determination of Reducing Sugars by the DNS Method [23,24]

The 3,5-Dinitrosalicylic acid (DNS) reagent was made by mixing in 5 g of dinitrosalicylic acid (Fluka Chemika, Fluka Chemie GmbH, Buchs, Switzerland) in 250 mL of distilled water at 80 °C in a water bath (Memmert, Memmert GmbH + Co., Äußere Rittersbacher Straße 38 D-91126 Schwabach, Germany). When this solution dropped to room temperature (25 °C), 100 mL of NaOH, 2 N (Fluka Chemika, Switzerland) and 150 g of sodium potassium tartrate (Fluka Chemika, Fluka Chemie GmbH, Buchs, Switzerland) were introduced, and the volume was completed with distilled water to 500 mL. The standard calibration curves were on glucose. According to the traditional method, 2 mL of dinitrosalicylic acid reagent and 0.1 mL of sample (must or wine), or distilled water (blank), were added to test tubes. The tubes were plunged in a water bath (100 °C) for 5 min and then cooled in cold water, while 7.9 mL of distilled water were introduced to each tube, resulting in the final reaction mixture. The addition of water while the tubes are plunged in cold water is performed to stop the reaction immediately. The optical density at 540 nm is read in the UV-visible spectrophotometer (JENWAY, Model: 7310, Serial No. 39756, JENWAY Limited, Staffordshire, UK).

2.2.16. Flavonoids

The flavonoid contents in the produced must and wine were determined utilizing spectrophotometric method [25]. The sample (20 µL) was added to 2 mL of 2% $AlCl_3$ solution dissolved in methanol. The samples were incubated for 1 h at room temperature (25 °C). The absorbance was assessed using the spectrophotometer at λ_{max} = 415 nm. The same steps were repeated for the standard solution of rutin, and the calibration line was built. Based on the determined absorbance, the concentration of flavonoids was read (mg/mL) on the calibration curve; and the flavonoid contents of must and wine were expressed in rutin equivalents (µg rutin/g of extract).

3. Results and Discussion

3.1. Ripening of Plantain

3.1.1. Percentage of Weight Loss

The weight loss evolution of giant horn plantain with ripening time is presented in Figure 3.

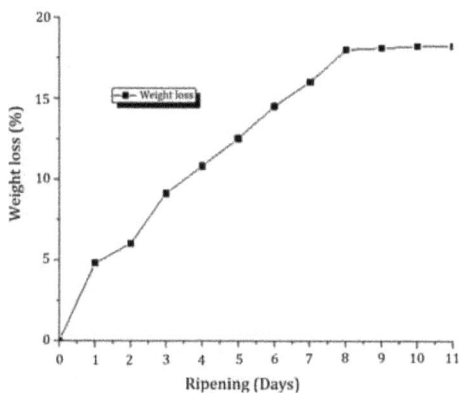

Figure 3. Weight loss (%) of giant horn plantain during ripening.

It was observed from the figure that the weight loss increased to reach a maximum weight loss of 18.2% after 10 days. The weight loss percentage obtained was higher than the 14.01% obtained after plantain ripening for 10 days [26–29]. The percent of the weight reduction of 'giant horn' plantain expanded constantly due to ripening because of a high storage temperature, ~25 °C. Like every chemical reaction, the plant metabolism increments with temperature. Since high energy was required to run the procedure, consequently, starch was changed over to sugar and utilized as energy. The most striking post-harvest compound change that happens amid the post-harvest ripening of plantain was the hydrolysis of starch and the amassing of sugars, like, sucrose, glucose and fructose [30,31], which were responsible for the sweetening of the product (as it matures). In plantain, this breakdown was slower and less complete and proceeds in over-ripe and senescent fruit [32]. The abundant energy created therefore from the respiration process [33,34] was dismissed from the tissue by the vaporization of water, which will in this way be transpired from the fruit, bringing on a weight reduction. A portion of humidity reduction through the peel could be seen through shrinkage on the peel.

3.1.2. Moisture Content

The moisture content of the ripe plantain was 63.4% ± 0.52%. This value was in the range of the value (61.3%) obtained in the literature [35]. The high moisture contents in the ripe plantain could be as a result of moisture transfer from the peel to the pulp. It was reported that the increase in moisture content of pulp occurred due to the increase in sugar content in the pulp as a result of starch hydrolysis to sugar [35,36].

3.2. Modelling and Optimizing of Juice Extraction

Table 5 presents the findings of the juice extraction carried out on the blanched plantain pulp. These results were later computed to develop statistical models for responses, amongst which were the yield and Brix. They are written as follows:

$$Y_{rdt} = 80.50 - 3.72x_1 + 0.81x_2 + 21.30x_3 + 0.32x_4 - 2.08x_1^2 + 2.77x_2^2 - 16.55x_3^2 + 4.11x_4^2 \\ -2.11x_1x_2 + 0.33x_1x_3 - 0.15x_1x_4 + 3.12x_2x_3 + 0.41x_2x_4 - 3.90x_3x_4 \tag{17}$$

$$Y_B = 11.025 + 0.810x_1 + 1.166x_2 - 12.852x_3 + 0.821x_4 - 0.26x_1^2 - 2.25x_2^2 + 9.31x_3^2 - 2.018x_4^2 \\ +1.01x_1x_3 - 0.23x_1x_4 - 1.11x_2x_3 - 0.61x_2x_4 + 0.41x_3x_4 \tag{18}$$

with: Y_B, Brix; Y_{rdt}, yield; x_1, extraction temperature; x_2, enzyme concentration; x_3, dilution factor and x_4, extraction time.

The R^2 values, AAD, B_f and A_f permitted the validation of the model. This was in accordance with the literature [18,37,38]. Factors modeled were linear (x_1, x_2, x_3 and x_4), quadratic (x_1^2, x_2^2, x_3^2 and x_4^2) and with interactions (x_1x_2, x_1x_3, x_1x_4, x_2x_3, x_2x_4 and x_3x_4). They were considered statistically significant or not if the probability (p) was ≤0.1 or ≥0.1, respectively (Table 6).

These models were the second-degree multivariable polynomial with interactions equations with the following characteristics: R^2 = 0.98 an, AAD = 0 for both, A_f = 1.016 and 1.065 and B_f = 1.000 and 0.997, respectively, for yield and Brix.

Table 5. Doehlert experimental design: coded variables, real variables and responses.

Number	Coded Variables				Real Variables				Responses					
									Yield (%)			Brix (°Bx)		
	x_1	x_2	x_3	x_4	X_1	X_2	X_3	X_4	Exp	Cal	Res	Exp	Cal	Res
1	1	0	0	0	75	2.55	2	12.25	73.10	74.70	−1.60	12.65	11.58	1.08
2	−1	0	0	0	25	2.55	2	12.25	83.75	82.14	1.61	8.90	9.96	−1.06
3	0.5	0.866	0	0	62.5	5	2	12.25	80.23	79.99	0.24	10.45	10.69	−0.24
4	−0.5	−0.866	0	0	37.5	0.1	2	12.25	82.08	82.30	−0.22	8.10	7.86	0.24
5	0.5	−0.866	0	0	62.5	0.1	2	12.25	81.73	80.41	1.32	8.15	8.67	−0.52
6	−0.5	0.866	0	0	37.5	5	2	12.25	84.23	85.53	−1.30	10.40	9.88	0.52
7	0.5	0.289	0.816	0	62.5	3.37	4	12.25	87.78	85.51	2.27	6.35	7.38	−1.03
8	−0.5	−0.289	−0.816	0	37.5	1.73	0	12.25	51.75	54.00	−2.25	27.90	26.87	1.03
9	0.5	−0.289	−0.816	0	62.5	1.73	0	12.25	49.68	50.62	−0.94	27.10	26.85	0.25
10	0	0.577	−0.816	0	50	4.18	0	12.25	51.48	52.02	−0.54	28.60	28.16	0.44
11	−0.5	0.289	0.816	0	37.5	3.37	4	12.25	90.53	89.57	0.96	5.50	5.74	−0.24
12	0	−0.577	0.816	0	50	0.92	4	12.25	86.40	85.85	0.55	5.40	5.84	−0.44
13	0.5	0.289	0.204	0.791	62.5	3.37	2.5	24	83.50	84.38	−0.88	8.75	8.53	0.22
14	−0.5	−0.289	−0.204	−0.791	37.5	1.73	1.5	0.5	79.33	78.44	0.89	10.80	11.01	−0.21
15	0.5	−0.289	−0.204	−0.791	62.5	1.73	1.5	0.5	76.63	75.38	1.25	11.00	11.79	−0.79
16	0	0.577	−0.204	−0.791	50	4.18	1.5	0.5	79.63	77.99	1.64	11.80	12.53	−0.73
17	0	0	0.612	−0.791	50	2.55	3.5	0.5	87.78	91.54	−3.76	6.25	4.54	1.71
18	−0.5	0.289	0.204	0.791	37.5	3.37	2.5	24	87.53	88.77	−1.24	8.50	7.70	0.80
19	0	−0.577	0.204	0.791	50	0.92	2.5	24	84.63	86.25	−1.62	7.95	7.22	0.73
20	0	0	−0.612	0.791	50	2.55	0.5	24	69.75	65.98	3.77	19.85	21.56	−1.71
21	0	0	0	0	50	2.55	2	12.25	80.83	80.50	0.33	11.35	11.03	0.32
22	0	0	0	0	50	2.55	2	12.25	80.28	80.50	−0.22	10.80	11.03	−0.23
23	0	0	0	0	50	2.55	2	12.25	80.76	80.50	0.26	11.00	11.03	−0.03
24	0	0	0	0	50	2.55	2	12.25	80.15	80.50	−0.35	10.95	11.03	−0.08
25	0	0	0	0	50	2.55	2	12.25	73.10	80.50	−7.40	12.65	11.03	1.63

Exp: Experimental value; Cal: Calculated value; Res: Residue (difference between experimental and calculated value)

Table 6. Estimation of the regression coefficients and variables' contributions for the yield and Brix of over-ripe giant horn plantain juice.

Effects	Coefficients		df	Sum of Squares		Mean Square		F-Value		p-Value	
	Yield	Brix		Yield	Brix	Yield	Brix	Yield	Brix	Yield	Brix
Constant	80.50	11.025	14	2949.24	1045.51	210.66	74.679	31.57	48.12	0.000	0.000
Linear			4	2338.06	838.24	584.51	209.559	87.61	135.03	0.000	0.000
x_1	−3.72	0.810	1	69.19	3.28	69.19	3.281	10.37	2.11	0.010	0.180
x_2	0.70	1.010	1	3.29	6.80	3.29	6.801	0.49	4.38	0.500	0.066
x_3	17.378	−10.487	1	2264.98	824.84	2264.98	824.837	339.47	531.49	0.000	0.000
x_4	0.256	0.650	1	0.52	3.38	0.52	3.377	0.08	2.18	0.786	0.174
Square			4	591.34	190.62	147.83	47.656	22.16	30.71	0.000	0.000
x_1^2	−2.08	−0.25	1	5.76	0.08	5.76	0.083	0.86	0.05	0.377	0.822
x_2^2	2.08	−1.685	1	10.24	6.75	10.24	6.750	1.53	4.35	0.247	0.067
x_3^2	−11.02	6.200	1	409.86	129.75	409.86	129.748	61.43	83.60	0.000	0.000
x_4^2	2.57	−1.262	1	28.17	6.80	28.17	6.799	4.22	4.38	0.070	0.066
Interactions			6	18.61	1.86	3.10	0.310	0.46	0.20	0.818	0.968
$x_1 x_2$	−1.82	0.00	1	3.33	0.00	3.33	0.000	0.50	0.00	0.498	1.000
$x_1 x_3$	0.27	0.82	1	0.07	0.61	0.07	0.612	0.01	0.39	0.923	0.545
$x_1 x_4$	−0.12	−0.18	1	0.01	0.03	0.01	0.029	0.00	0.02	0.966	0.894
$x_2 x_3$	2.20	−0.79	1	5.83	0.74	5.83	0.742	0.87	0.48	0.374	0.507
$x_2 x_4$	0.28	−0.41	1	0.09	0.20	0.09	0.204	0.01	0.13	0.909	0.725
$x_3 x_4$	−2.52	0.27	1	8.45	0.10	8.45	0.095	1.27	0.06	0.290	0.810
Residual			9	60.05	13.97	6.67	1.552				
total			23	3009.28	1059.47						

3.2.1. Impact of Extraction Temperature

The impact of extraction temperature was only significant ($p = 0.010$, Table 6) on the extraction yield. In fact, that temperature contributed to reducing the yield from 53.52% at 25 °C to 49.43% at 75 °C (Figure 4) when fixing the enzyme concentration at 0.1%, the dilution factor at zero (no dilution) and the extraction time at 0.5 h. It was found that for an average of several hybrids of plantains, an amount of water was absorbed after 30 min cooking [39,40]. When heated in excess water with a sufficient amount of heat, native starch granules undergo irreversible phase transition, called gelatinization [41]. The highly-ordered structure becomes disordered. Gelatinization of starch has been broadly defined as an irreversible endothermic transition, the "breakage of the molecular order", breaking the hydrogen bonds in the starch granule. Irreversible changes are observed: water gain, swelling of the granule, crystalline melting, loss of birefringence, the solubilization of the starch and the increase in viscosity [42]. During the heat treatment, the water enters in the first place in the amorphous regions, which swell and transmit disturbing forces in the crystalline regions [43,44]. All of this results in the reduction of yield extraction.

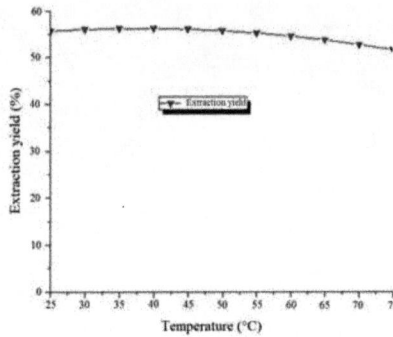

Figure 4. Evolution of extraction yield with temperature.

3.2.2. Impact of Enzyme Concentration

The impact of enzyme concentration was this time only significant ($p = 0.066$, Table 6) on the Brix. In fact, the enzyme concentration contributed to increase the Brix from 21.73 °Bx at 0.1% to 26.16 °Bx at 5% (Figure 5) when fixing the extraction temperature at 25 °C, the dilution factor at zero (no dilution) and the extraction time at 0.5 h. The use of multiple enzymes increased the TSS (total soluble solids) content of juice. The enzymatic effect permitted the cell walls to be more permeable, which generated the extraction of total soluble solids from the plantain puree. The presence of polysaccharide-degrading enzymes (amyloglucosidase, α-amylase, pectinase and bioglucanase activities) in the preparations was then linked to a higher degree of tissue breakdown, discharging more compounds, such as sugars [45], which contribute to total soluble solids.

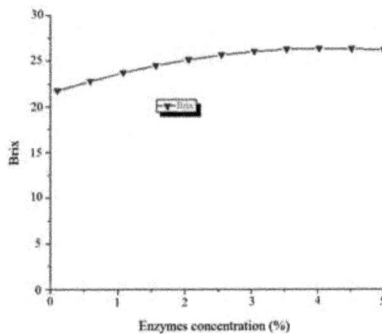

Figure 5. Evolution of Brix with enzyme concentration.

3.2.3. Impact of the Dilution Factor

The effect of the dilution factor was significant ($p = 0.000$ for both; Table 6) on the increase of extraction yield and the decrease of Brix.

The increase for extraction yield started from 53.52% at zero (no dilution) to 92.77% at four, while the decrease of Brix (Figure 6) started from 21.74 °Bx at zero (no dilution) to 0.15 °Bx at four; all this when fixing the temperature at 25 °C, the enzyme concentration at 0.1% and the extraction time at 0.5 h.

The increase of water during the extraction process contributed to the increase of extraction yield because at the fixed temperature of 25 °C, the starch granules were not gelatinized, and therefore, the more water was added during the process, the more it was recovered during centrifugation to obtain the juice; while, at the same time, the increase of water when proceeding to extraction contributed to diluting the total soluble solids of that juice, which had as the impact the reduction of the Brix, which expresses the concentration (in percentage) of total soluble solids in the juice.

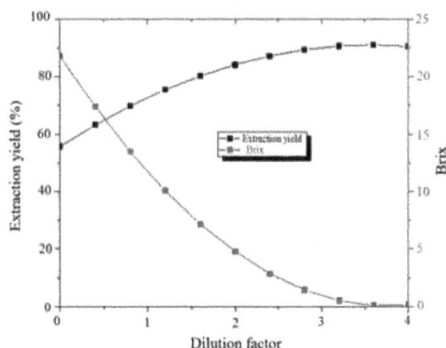

Figure 6. Evolution of extraction yield and Brix with the dilution factor.

3.2.4. Optimization of Juice Extraction

Production of the juice took globally into account the compromise between the extraction yield and the physicochemical composition of the juice extracted, since the extraction procedure could influence the composition.

When trying to maximize the responses singularly, different optimal combinations were obtained. For the Brix, the maximal value obtained was 28.19 °Bx for an optimal combination of 48.74 °C for temperature, 3.81% for enzyme concentration, zero (no dilution) for the dilution factor and 13.08 h for the extraction time; while the maximization of the extraction yield gave 99.47% using 25 °C for the temperature, 5% for the enzyme concentration, four for the dilution factor and 0.5 h for the extraction time. From the two singulars response maximization, it was obvious that a compromise should be found. For that purpose, an extraction yield $\geq$70% and $10 \leq$ Brix ≤ 15 °Bx were fixed as "good" conditions. That choice was made because of the aim of producing a wine (low Brix) and to have at the same time enough juice volume (high extraction yield). After superimposing the graphs for Brix and extraction yield (Figure 7), it was obtained that the area that could permit reaching the target (extraction yield $\geq$ 70% and $10 \leq$ Brix ≤ 15 °Bx) was within the interest zone highlighted (that zone was obtained after fixing the temperature at 25 °C and the extraction time at 24 h).

The optimal condition in the highlighted area was assessed as follows: temperature, 25 °C; enzyme concentration, 5%; dilution ratio, 1.10; and extraction time, 24 h. That optimal combination gave respectively for Brix 14.99 °Bx and for the extraction yield 80.22%.

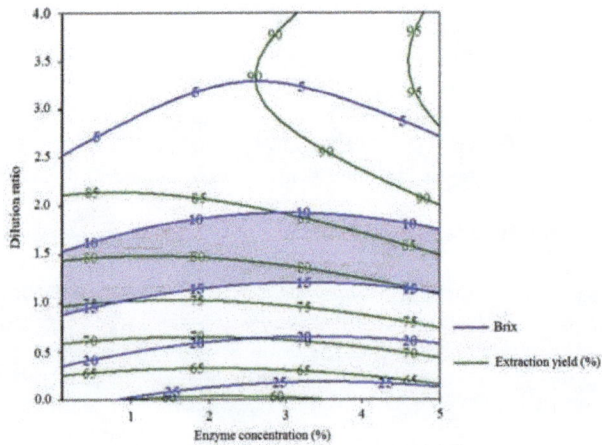

Figure 7. Interest zone for giant horn plantain juice extraction.

3.3. Evolution of Turbidity during Fermentation

Figure 8 shows the evolution of turbidity during the fermentation of over-ripe giant horn plantain juice extraction using optimal conditions and during the first fermentation. Turbidity increased from 144.33 ± 0.57 NTU (the initial day) to 1901.33 ± 26.31 NTU (the first day). After that first day, the turbidity decreased till the value of 88.66 ± 0.57 NTU. The high turbidity observed the first day of fermentation could be due to the fact that the yeast metabolized the sugars, and that produced a high volume of CO_2.

Gas escaping created turbulence in the medium and dispersing yeast. The progressive reduction of turbidity could be linked to the reduction of CO_2 production and, at the same time, the beginning of yeast and other solids' decantation. After 96 h, the drop of turbidity was drastic. This could be explained by a significant reduction of CO_2 gas and, therefore, a more efficient decantation.

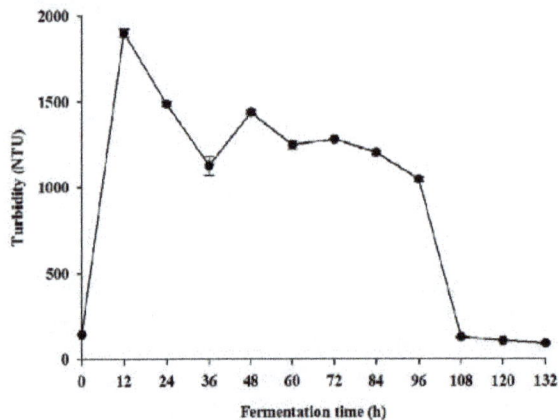

Figure 8. Evolution of turbidity during fermentation of over-ripe giant horn plantain juice.

3.4. Physicochemical Characteristics of Must and Wine

Some physicochemical characteristics of the must and the wine (Figure 9) at the end of the 90 days of the aging process are presented in Table 7.

After extraction in optimal conditions, which gives 15.1 ± 0.42 °Bx, the juice was then pasteurized to prevent fermentation starting prematurely by indigenous yeasts and bacteria.

Figure 9. Over-ripe giant horn plantain wine.

From Table 7, it was observed that the pH dropped from 4.26 ± 0.01 in must to 4.12 ± 0.21 in wine. During the alcoholic fermentation, the pH of the must was constantly changing. Organic acids consumed or produced undergo dissociation and release hydrogen ions in the fermentation medium, thus influencing the pH. All hydrogen ions did not come, however, from dissociation of organic acids. Some came from the assimilation of nitrogen source, including ammonium ions, probably amino acids, which could contribute to the pH drop by amino acid assimilation of a positive charge. Alcohol was the main product of fermentation and was also involved in the change in pH. It acted on the dissociation constants of organic acids, on the density and the dielectric constant of the solvent and, thus, indirectly on the pH [46].

Table 7. Physicochemical analysis of must and wine.

Analysis	Must	Wine
pH	4.26 ± 0.01	4.12 ± 0.21
Total soluble sugar (°Bx)	16.06 ± 0.58	1.5 ± 0.10
Titratable acidity (g/L tartaric acid)	4.4 ± 0.09	8.19 ± 0.03
Flavonoids (μg rutin/g of sample)	0.4 ± 0.01	4 ± 0.12
Polyphenols (mg GAE/100 g of sample)	7.70 ± 0.99	17.01 ± 0.34
Color (EBC: European Brewery Convention unit)	23.67 ± 1.50	20.03 ± 0.04
Total sugars (g/100 mL)	9.61 ± 0.14	3.23 ± 0.09
Free amino nitrogen (mg/L)	138.94 ± 5.23	78.77 ± 2.75

For the total soluble solids, a decrease of Brix from 16.06 ± 0.58 °Bx to 1.5 ± 0.10 °Bx was observed (Table 7). Yeasts are facultative aerobic organisms. When oxygen was available, glucose was metabolized aerobically. In the absence of air, yeast must go against this by alcoholic fermentation. As it produced less energy than aerobic respiration, the need for glucose increased significantly. This phenomenon is named the Pasteur effect. Due to the limited production of energy, yeast multiplied in the absence of air much less quickly than in its presence. In addition, the ethanol produced acted as a cellular poison. The need for glucose by the yeast contributed then to the decrease of sugar in the medium.

Furthermore, the fact that the residual sugar content was 1.5 ± 0.10 °Bx could then be explained as said earlier. The titratable acidity obtained was 4.4 ± 0.09 g/L tartaric acid for the must and increased to 8.19 ± 0.03 g/L tartaric acid for wine (Table 7). Organic acids are essential elements of the constitution of musts and wines, their qualities and their defects. Nature and concentration regulate the acid-base balance and, therefore, control the acid taste of wine. During the fermentation, sugar consumption by the yeast (glycolysis) results in the formation of typical organic acid fermentation [46].

The flavonoid content of the giant horn plantain must and wine is shown in Table 7. The flavonoid content of must and wine samples was from 0.4 ± 0.01 µg rutin/g–4 ± 0.12 µg rutin/g, respectively. The augmentation in flavonoid content in giant horn plantain wine as a result of fermentation could be due to the increase in acidity during fermentation, which involves unchaining bound flavonoid components and making them more available [47].

The total phenolic content of the giant horn plantain must and wine expressed as mg gallic acid equivalent (GAE) per 100 g sample is shown in Table 7. Phenolic content for must and wine samples analyzed in this study ranged from 7.70 ± 0.99 mg GAE/100 g to 17.01 ± 0.34 mg GAE/100 g, respectively. In natural medium, phenolic compounds are linked with sugar, which decreases their disponibility. During fermentation, proteases hydrolyze complexes of phenolics into soluble-free phenols and other simpler and biologically more active ones that are readily absorbed [48,49].

From Table 7, a decrease in color from must to wine was also observed. That color level was 23.67 ± 1.50 EBC for must and 20.03 ± 0.04 EBC for wine. This could be explained by the yeast, which has been shown to absorb anthocyanins onto its cell walls and cause color loss [50].

After analyzing free amino nitrogen (FAN) of the must and at the end of the fermentation (wine) in Table 7, a drop from 138.94 ± 5.23 mg/L (must) to 78.77 ± 2.75 mg/L (wine) was observed. Generally, during alcoholic fermentation, ammonium ions and most amino acids are completely consumed after the first 50 h of fermentation [51,52]. It is also known that amino acids are involved in the synthesis of higher alcohols by the yeast mechanism. Thus, the more a must is concentrated in available nitrogen, a larger amount is consumed, as pointed out by the literature [53].

4. Conclusions

This study was aimed at valorizing over-ripe plantains in order to reduce post-harvest losses by processing them into wine owing to 35% of the total plantains produced being wasted as post-harvest losses. Mathematical models for the must extraction were successfully developed in order to predict the appropriate conditions for optimal yield and Brix. A compromise of the optimum processing conditions for both responses (extraction yield and Brix) was determined as 80.22% and 14.99 °Bx, respectively. The must and wine produced were characterized for pH, turbidity, total soluble sugars, total polyphenols, free amino nitrogen, color, total sugars, flavonoids and titratable acidity and tend to have suitable properties, as well. It has been determined that over-ripe plantains that may be available in bulk should not be wasted, as they can be processed with minimal problems to produce safe and quality alcoholic beverages (wines).

Acknowledgments: The authors gratefully acknowledge the Department of Process Engineering, National School of Agro-Industrial Sciences (ENSAI) of the University of Ngaoundere (Cameroon) for providing necessary facilities for the successful completion of this research work.

Author Contributions: W. C. Makebe, performed the experiments. Z. S. C. Desobgo, analyzed the data; contributed reagents/materials/analysis tools; wrote the paper. E. J. Nso, contributed reagents/materials/analysis tools.

Conflicts of Interest: The authors declare no conflicts of interest.

Abbreviations

ANOVA	Analysis of variance
AAD	Absolute average deviation
AAE	Ascorbic acid equivalent
A_f	Exactitude factor
B_f	Bias factor
°Bx	Degree Brix
Cal	Calculated value from the model
DM	Dry matter
DF	Degree of freedom
EBC	European Brewing Convention
Exp	Experimental value
GAE	Gallic acid equivalent
N	Number of experiments
n_0	number of points repeated independently
NTU	Nephelometric turbidity unit
P	Number of model coefficients
R^2	Coefficient of determination
Res	Residue (Difference between experimental and calculated values)
RSM	Response surface methodology
SS	Sum of squares
SSE	Sum square of error
SSE_1	Sum of the squares of the experimental errors (at the repeated points)
SSE_2	Sum of the squares of the deviations (at other points) or validity.
SSR	Sum of squares due to regression
TSS	Total sum of squares

References

1. Olurunda, A.O. Recent advances in post-harvest technology of banana and plantains in Africa. *Acta Hort.* **2000**, *540*, 517–527. [CrossRef]
2. Essilfie, G. *Analysis of the Post-Harvest Knowledge System in Ghana: Case Study of Cassava*; CTA: Wageningen, The Netherlands, 2014; p. 10.
3. Food and Agriculture Organization (FAO). *Faostat Database Collections*; Food and Agriculture Organization of the United Nations: Rome, Italy, 2014.
4. Adeniji, T.A.; Tenkouano, A.; Ezurike, J.N.; Ariyo, C.O.; Vroh-Bi, I. Value-adding post harvest processing of cooking bananas (*Musa spp.* Aab and abb genome groups). *Afr. J. Biotechnol.* **2010**, *9*, 9135–9141.
5. Hailu, M.; Workneh, T.S.; Belew, D. Review on postharvest technology of banana fruit. *Afr. J. Biotechnol.* **2013**, *12*, 635–647.
6. Strosse, H.; Schoofs, H.; Panis, B.; André, E.; Reyniers, K.; Swennen, R. Development of embryogenic cell suspensions from shoot meristematic tissue in bananas and plantains (*Musa* spp.). *Plant Sci.* **2006**, *170*, 104–112. [CrossRef]
7. Fleet, G.H. *Wine: Microbiology and Biotechnology*; Harwood Academic Publishers: London, UK, 1993; p. 130.
8. Webb, A.D. The science of making wine. *Am. Sci.* **1984**, *72*, 360–367.
9. Isitua, C.C.; Ibeh, I.N. Novel method of wine production from banana (*Musa acuminata*) and pineapple (ananas comosus) wastes. *Afr. J. Biotechnol.* **2010**, *9*, 7521–7524. [CrossRef]
10. Jay, J.M. *Modern Food Microbiology*, 5th ed.; Chapman and Hall: New York, NY, USA, 1996; p. 212.
11. Okafor, N. *Modern Industrial Microbiology and Biotechnology*; Science Publishers: Enfield, NH, USA, 2007; p. 530.
12. Kotecha, M.P.; Adsule, R.N.; Kadam, S.S. Preparation of wine from over-ripe banana fruits. *Beverage Food World* **1994**, *21*, 28–29.
13. Akingbala, J.O.; Oguntimein, G.B.; Olunlade, B.A.; Aina, J.O. Effects of pasteurization and packaging on properties of wine from over-ripe mango (mangifera indica) and banana (musaacuminata) juices. *Trop. Sci.* **1992**, *34*, 345–352.
14. Goupy, J.; Creighton, L. *Introduction Aux Plans D'expériences*, 3rd ed.; Dunod: Paris, France, 2006; p. 325.

15. Ekorong, A.A.J.F.; Zomegni, G.; Desobgo, Z.S.C.; Ndjouenkeu, R. Optimization of drying parameters for mango seed kernels using central composite design. *Bioresour. Bioprocess* **2015**, *2*, 1–9. [CrossRef]

16. Joglekar, A.M.; May, A.T. Product excellence through design of experiments. *Cereal Foods World* **1987**, *32*, 857–868.

17. Baş, D.; Boyac, I.H. Modeling and optimization i: Usability of response surface methodology. *J. Food Eng.* **2007**, *78*, 836–845. [CrossRef]

18. Dalgaard, P.; Jorgensen, L.V. Predicted and observed growth of listeria monocytogenes in seafood challenge tests and in naturally contaminated cold smoked salmon. *Int. J. Food Microbiol.* **1998**, *40*, 105–115. [CrossRef]

19. ISO 2173:2003. *Fruit and Vegetable Products—Determination of Soluble solids—Refractometric Method*; International Organization of Standardization (ISO): Geneva, Switzerland, 2003.

20. Marigo, G. Méthode de fractionnement et d'estimation des composés phénoliques chez les végétaux. *Analysis* **1973**, *2*, 106–110.

21. Association Française de Normalisation (AFNOR). *Recueil Des Normes Françaises Des Produits Dérivés Des Fruits Et Légumes*, 1st ed.; Association Française de Normalisation: Paris, France, 1982.

22. Analytica-EBC. *European Brewery Convention*; Fachverlag Hans Carl: Nürnberg, Germany, 1998.

23. Fisher, E.H.; Stein, E.A. Dns colorimetric determination of available carbohydrates in foods. *Biochem. Prep.* **1961**, *8*, 30–37.

24. Fisher, E.H.; Stein, E.A. *Enzymes*, 2nd ed.; Academic Press: New York, NY, USA, 1960; Volume 4, p. 343.

25. Quettier-Deleu, C.; Gressier, B.; Vasseur, J.; Dine, T.; Brunet, J.; Luyck, M.; Cazin, M.; Cazin, J.C.; Bailleul, F.; Trotin, F. Phenolic compounds and antioxidant activities of buckwheat (*Fagopyrum esculentum* moench) hulls and flour. *J. Ethnopharmacol.* **2000**, *72*, 35–40. [CrossRef]

26. Oluwalana, I.B. Fruit cultivation and processing improvement in nigeria. *Agric. J.* **2006**, *1*, 307–310.

27. Oluwalana, I.B. Minimizing fruit wastages in nigeria. *Int. J. Agric. Food Sci.* **2010**, *1*, 77–87.

28. Oluwalana, I.B.; Oluwamukomi, M.O. Changes in qualities of ripening plantain fruits stored at tropical ambient conditions. *Int. J. Agric. Food Sci.* **2010**, *1*, 203–207.

29. Oluwalana, I.B.; Oluwamukomi, M.O.; Olajide, S.T. Physicochemical changes in ripening plantain stored at tropical ambient conditions. *J. Food Technol.* **2006**, *4*, 253–254.

30. Lœsecke, H.V. *Bananas*, 2nd ed.; InterScience: New York, NY, USA, 1950.

31. Palmer, J.K. The banana. In *The Biochemistry of Fruits and Their Products*; Hulme, A.C., Ed.; Academic Press: London, UK, 1971; Volume 2, pp. 65–105.

32. Marriott, J.; Robinson, M.; Karikari, S.K. Starch and sugar transformation during ripening of plantains and bananas. *Trop. Sci.* **1981**, *32*, 1021–1026. [CrossRef]

33. George, J.B.; Marriott, J. The effect of humidity in plantain ripening. *Sci. Holtic.* **1983**, *5*, 37–48. [CrossRef]

34. Pantastico, B.E. *Postharvest Physiology, Handling and Utilization of Tropical and Subtropical Fruits and Vegetables*; AVI Publishing Co. Inc.: Westport, CT, USA, 1975.

35. Egbebi, A.O.; Bademosi, T.A. Chemical compositions of ripe and unripe banana and plantain. *Int. J. Trop. Med. Public Health* **2012**, *1*, 1–5.

36. Patil, S.K.; Shanmugasundaram, S. Physicochemical changes during ripening of monthan banana. *Int. J. Technol. Enhanc. Emerg. Eng. Res.* **2015**, *3*, 18–21.

37. Ross, T. Indices for performance evaluation of predictive models in food microbiology. *J. Appl. Bacteriol.* **1996**, *81*, 501–508. [PubMed]

38. Baranyi, J.; Pin, C.; Ross, T. Validating and comparing predictive models. *Int. J. Food Microbiol.* **1999**, *48*, 159–166. [CrossRef]

39. Dadzie, B.K. Cooking qualities of black sigatoka resistant plantain hybrids. *Infomusa* **1995**, *4*, 7–9.

40. Ngalani, J.A.; Tchango-Tchango, J. Cooking qualities and physicochemical changes during ripening in some banana and plantain hybrids and cultivars. *Acta Hortic.* **1998**, *490*, 571–576. [CrossRef]

41. Wang, S.; Copeland, L. Molecular disassembly of starch granules during gelatinization and its effect on starch digestibility: A review. *Food Funct.* **2013**, *4*, 1564–1580. [CrossRef] [PubMed]

42. Donovan, J.W. Phase transitions of the starch–water system. *Biopolymers* **1979**, *18*, 263–275. [CrossRef]

43. BeMiller, J.N. Pasting, paste, and gel properties of starch–hydrocolloid combinations. *Carbohydr. Polym.* **2011**, *86*, 386–423. [CrossRef]

44. Slade, L.; Levine, H. Non-equilibrium melting of native granular starch: Part i. Temperature location of the glass transition associated with gelatinization of a-type cereal starches. *Carbohydr. Polym.* **1988**, *8*, 183–208. [CrossRef]

45. Sreenath, H.K.; Frey, M.D.; Scherz, H.; Radola, B.J. Degradation of a washed carrot preparation by cellulases and pectinases. *Biotechnol. Bioeng.* **1984**, *26*, 788–796. [CrossRef] [PubMed]

46. Akin, H. *Evolution Du Ph Pendant La Fermentation Alcoolique De Moûts De Raisins: Modélisation Et Interprétation Métabolique*; Institut National Polytechnique de Toulouse Toulouse: Toulouse, France, 2008.

47. Adetuyi, F.O.; Ibrahim, T.A. Effect of fermentation time on the phenolic, flavonoid and vitamin C contents and antioxidant activities of okra (*abelmoschus esculentus*) seeds. *Niger. Food J.* **2014**, *32*, 128–137. [CrossRef]

48. Shrestha, A.K.; Dahal, N.R.; Ndungustse, V. Bacillus fermentation of soybean: A review. *J. Food Sci. Technol. Nepal* **2010**, *6*, 1–9. [CrossRef]

49. Ademiluyi, A.O.; Oboh, G. Antioxidant properties of condiment produced from fermented bambara groundnut (*Vigna subterranean* L. Verdc). *J. Food Biochem.* **2011**, *35*, 1145–1160. [CrossRef]

50. Morata, A.; Gomez-Cordoves, M.C.; Colomo, B.; Suarez, J.A. Pyruvic acid and acetaldehyde production by different strains of saccharomyces: Relationship with vitisin a and b formation in red wines. *J. Agric. Food Chem.* **2003**, *51*, 7402–7409. [CrossRef] [PubMed]

51. Jiranek, V.; Langridge, P.; Henschke, P.A. Amino acid and ammonium utilization by saccharomyces cerevisiae wine yeasts from a chemically defined medium. *Am. J. Enol. Vitic.* **1995**, *46*, 75–83.

52. Beltran, G.; Novo, M.; Rozès, N.; Mas, A.; Guillamón, J.M. Nitrogen catabolite repression in saccharomyces cerevisiae during wine fermentations. *FEMS Yeast Res.* **2004**, *4*, 625–632. [CrossRef] [PubMed]

53. Taillandier, P.; Ramon-Portugal, F.; Fuster, A.; Strehaiano, P. Effect of ammonium concentration on alcoholic fermentation kinetics by wine yeasts for high sugar content. *Food Microbiol.* **2007**, *24*, 95–100. [CrossRef] [PubMed]

beverages

MDPI

Article

Physicochemical Stability, Antioxidant Activity, and Acceptance of Beet and Orange Mixed Juice during Refrigerated Storage

Maria Rita A. Porto [1], Vivian S. Okina [1], Tatiana C. Pimentel [2] and Sandra Helena Prudencio [1,*

[1]   Departamento de Ciência e Tecnologia de Alimentos, Centro de Ciências Agrárias, Universidade Estadual de Londrina, Londrina 10011, Paraná, Brazil; mariaritaalaniz@hotmail.com (M.R.A.P.); viviansayuri@gmail.com (V.S.O.)

[2]   Instituto Federal do Paraná, Rua José Felipe Tequinha 1400, 87703-536 Paranavaí, Paraná, Brazil; tatiana.pimentel@ifpr.edu.br

Academic Editor: António Manuel Jordão
Received: 23 May 2017; Accepted: 13 July 2017; Published: 18 July 2017

Abstract: The objective of this study was to mix beet juice and orange juice in two proportions (1:1 and 1:2 v/v), evaluate their physicochemical stability and antioxidant activity during storage (4 °C for 30 days), and evaluate their acceptance by consumers. Beet juice (with or without pasteurization) and pasteurized orange juice were used as controls. The presence of orange juice contributed to the pH, betacyanin, betaxanthin, and antioxidant capacity stabilities during storage, whereas the presence of beet improved the color stability. The mixed juices showed high total phenolic compounds (484–485 µg gallic acid/mL), DPPH scavenging capacity (2083–1930 µg Trolox/mL), and ABTS (1854–1840 µg Trolox/mL), as well as better sensory acceptance than the pasteurized beet juice. However, the mixed juices had a more significant reduction in the ascorbic acid content (completely lost at 15 days of storage) than the pasteurized orange juice (25% reduction at 30 days). The beet and orange mixed juice is an alternative functional beverage that can contribute to an increase in the consumption of beet and orange.

Keywords: phenolics; betacyanin; ascorbic acid; *Beta vulgaris* L.; *Citrus X sinensis*

---

1. Introduction

Beet (*Beta vulgaris* L.) is one of the most produced green vegetables in the world [1]. Beet juice is a convenient alternative for vegetal consumption, and contains beneficial compounds for health, such as potassium, magnesium, folic acid, iron, zinc, calcium, phosphate, sodium, niacin, biotin, the B6 vitamin, and soluble fiber. Furthermore, beet juice is a rich source of polyphenolic compounds, which are biologically available antioxidants [2].

The intense and attractive color of the beet is due to the presence of hydrosoluble pigments called betalains. The betalains are derived from betalamic acid and divided into two sub-classes: betacyanin (purple-red pigments) and betaxanthins (yellow-orange pigments) [3]. These pigments are important antioxidant substances, and present antimicrobial and anti-inflammatory activities, the inhibition of lipid peroxidation, an increased resistance to oxidation of low-density lipoproteins, and chemo-preventive effects [3–5].

Orange juice (*Citrus sinensis*) is responsible for half of the juice produced in the world, and it has great acceptability because of its flavor. Orange juice is a source of ascorbic acid, a nutrient that has a vitaminic action, and is a co-factor in several physiological processes, such as lysine and proline hydroxylation, the synthesis of collagen and other proteins of the conjunctive tissue, the synthesis of the norepinephrine and adrenal hormones, the activation of peptide hormones, and the synthesis of carnitine. Ascorbic acid also acts as an antioxidant, and facilitates iron absorption by the intestines and the maintenance of ferrous ions in blood plasma [6–8].

According to recommendations by the World Health Organization (WHO) and the Food and Agriculture Organization of the United Nations (FAO), the minimum consumption of fruit and green vegetables for adults must be of 400 g/day (five or more daily portions) [9]. However, according to data from the Chronic Diseases and Risk Factors Surveillance, only 20.2% of Brazilians eat the recommended daily amount [10].

One of the alternatives to redress the balance between the real and the recommended ingestions in a diet is the consumption of functional beverages rich in fruit and/or green vegetables. These products can also be a source of antioxidants, provided that their functional components have been preserved, and their commercialization will be effective if they are widely accepted by consumers [11].

Therefore, the combination of beet juice, rich in antioxidant compounds, with orange juice, highly acceptable sensorially and rich in ascorbic acid and polyphenols, results in an interesting beverage. As far as the authors know, there are no studies concerning the development of beet and orange mixed juices and their stability during refrigerated storage. Therefore, the objective of this study was to evaluate the acceptance as well as the physicochemical stability and antioxidant activity of beet and orange mixed juices (1:1 or 1:2 v/v) during refrigerated storage for 30 days.

2. Materials and Methods

2.1. Preparation of the Juices

Five formulations of juices were prepared: non-pasteurized beet juice (BJ); pasteurized beet juice (PBJ); pasteurized orange juice (POJ); beet and orange mixed juices, 1:1 (BOMJ 1:1); and beet and orange mixed juices, 1:2 (BOMJ 1:2).

Fresh beets (*Beta vulgaris* L.) cv. Early Wonder were obtained from Central de Abastecimento do Paraná S.A. (CEASA) in the city of Londrina, Brazil. The beets were washed in running water, sanitized (150 mg/L Dinamica® fruit disinfectant with sodium hypochlorite for 30 min), and manually peeled and crushed using a fruit processor (Mondial®, Brazil), originating the beet juice. To obtain the orange juice, concentrated orange juice (66° Brix) was diluted with sterilized water in the proportion of 1:6 (v/v), according to the manufacturer's recommendation, until the soluble solids value was close to 12° Brix. The concentrated orange juice was sugarless and preservative-free.

The beet and orange mixed juices were prepared from beet juice and the diluted orange juice, mixing them in two proportions, 1:1 (BOMPJ 1:1) or 1:2 (BOMPJ 1:2) (v/v), respectively. The proportions of the beet and orange in the mixed juices were based on a preliminary analysis in order to obtain a product with suitable phenolic compounds and physicochemical and sensory characteristics. The juices (PBJ, POJ, BOMJ 1:1, and BOMJ 1:2) were placed in transparent glass flasks (Farma®, Brazil), and pasteurized at 70 °C for 30 min in a water bath. The BJ juice was not submitted to the heat treatment.

The juices were stored at 4 °C for 30 days, which is the shelf life of commercial pasteurized Brazilian juices. The products were evaluated on days 0, 5, 10, 15, and 30 of storage.

2.2. Physicochemical Analysis

The pH determination was carried out by a calibrated digital potentiometer (Tecnal, Brazil). The titratable acidity was measured by titration with a 0.01 M NaOH (Anidrol®, Brazil) solution until pH 8.2, and expressed as a citric acid percentage [12].

A colorimeter (Minolta®, model CR400, Osaka, Japan) was used for the assessment of the color parameters values (L*, a*, b*, and C (Chroma)). From parameters L*, a*, and b*, the total difference was calculated (ΔE^*ab) between each sample at a determined storage time, and the sample at the initial time of storage, as follows: $\Delta E^*ab = [(\Delta L^*)^2 + (\Delta a^*)^2 + (\Delta b^*)^2]^{1/2}$.

2.3. Ascorbic Acid Determination by High Performance Liquid Chromatography (HPLC)

The ascorbic acid content was determined according to Souza et al. [13], with modifications. The juice samples that contained orange juice (POJ, BOMJ 1:1, and BOMJ 1:2) were dissolved in 3% metaphosphoric acid and filtered in a PVDF hydrophilic membrane (pores of 0.22 µm) (Millex-GV, Millipore, USA). The chromatograph system (Shimadzu, Japan) consisted of two pumps (model LC-10AD), a Rheodyne injecting valve with a sample handle of 20 µL, a column oven (model CTO-20A), a UV/visible spectrophotometric detector (model SPD-10A), an interface (model CBM-101), the CLASS-CR10 program, version 1.2 (Shimadzu, Japan), and a Spherisorb ODS1 (4.6 × 250 mm, 5 µm) column (Waters®, Ireland). The analysis was conducted at room temperature using an isocratic elution (0.5 mL/min) with a sulfuric acid solution (pH 2.5) as the mobile phase. The ascorbic acid detection was done at 254 nm. The identification was based on retention times and co-elution with standards, and the quantification by external standardization using a calibration curve with six points (measured in duplicate) in the concentration range of 5 to 50 µg/mL.

2.4. Betalains Determination

For the extraction of the betalains, sample aliquots (except POJ) of 1 mL were diluted in 30 mL distilled water by manual agitation for 10 s, and the mix was centrifuged (25 °C) at 6000× g for 20 min (Eppendorf centrifuge 5804R, Germany). The betaxanthin and betacyanin content in the supernatants were analyzed through absorbance readings at 476 and 538 nm in a UV/VIS spectrophotometer (Biochrom Libra S22, UK). A reading at 600 nm was used to correct possible impurities (turbidity) according to the methods of Stintzing, Schieber, and Carle [14], with some modifications. The betalain content (BC), expressed in mg/mL of juice, was calculated from the equation = $[(A \times FD \times MM \times 100)/(\varepsilon \times I)]$, where A is the difference between maximum absorption (476 or 538 nm) and the correction absorption at 600 nm, FD is the dilution factor, and I is the length of the cuvette path (1 cm). For the quantification of betacyanin and the betaxanthins, the molecular masses (MM) and molar absorptivity (ε) were: MM = 550 g/mol, ε = 60,000 L/mol cm in H2O and MM = 308 g/mol, ε = 48,000 L/mol cm in H2O, respectively.

2.5. Total Phenolic Compounds Content Determination

The phenolic compounds were determined by the Folin–Ciocalteau method, as described by Singleton, Orthofer, and Lamuela-Raventos [15]. An aliquot (0.2 mL) of sample was added to 1.5 mL of Folin–Ciocalteau aqueous solution (1:10 v/v). The mix was left to stabilize for 5 min, and then mixed with 1.5 mL of a sodium carbonate solution (60 g/L). Then, it was incubated at room temperature for 90 min and the absorbance of the mix was read at 725 nm in a UV-VIS spectrophotometer, using the respective solvent as the blank. The results were expressed in µg of gallic acid equivalents (µg EAG/100 mL).

2.6. Antioxidant Activity Evaluated by the DPPH Free Radical Scavenging Method

The total antioxidant capacity was evaluated by the DPPH method according to Brand-Williams, Cuvelier, and Berset [16], with some modifications. In a test tube were mixed 1 mL of 100 mM sodium acetate buffer (pH 5.5), 1 mL of absolute ethanol, 0.5 mL of an ethanolic solution of DPPH (250 µmol/L), and 50 µL of each sample. The test tubes were kept in the dark and at room temperature for 30 min, and the absorbance was read at 517 nm in a UV-VIS spectrophotometer. The antioxidant activity quantification was realized by a Trolox standard curve, and the results expressed in µmol of Trolox/mL.

2.7. ABTS+ Cation Radical Scavenging Activity

The antioxidant activity was also analyzed by the sample capacity to eliminate ABTS free radicals, using the modified methodology reported by Ozgen et al. [17]. When combined with an oxidant (potassium persulphate 2.45 mM), the ABTS (7 mM in phosphate tampon 20 mM, pH 7.4) reacts to create a dark greenish-blue stable radical solution after 12–16 h of incubation in the dark. The solution was diluted up to an absorbance of 0.7 ± 0.01 in 730 nm to form the reagent test. Reactional mixes with 10 μL of sample and 4 mL of reagent test were incubated at room temperature for 6 min. The antioxidant activity was calculated in relation to the Trolox standard curve and expressed in μmol of Trolox/mL.

2.8. Sensory Analysis

The sensory analysis was conducted in the juices that had beet in the formulation (PBJ, BOMJ 1:1, and BOMJ 1:2) on the 5th day of storage. The pure orange juice (POJ) was not evaluated for acceptance because it is known that it has great acceptability, and the objective was to produce a juice with the phenolic compounds provided by beet. The acceptance test of attributes (color, aroma, flavor, and overall impression) was conducted with a 9-point hedonic scale (1 = disliked extremely and 9 = liked extremely). The formulations were coded with 3-digit numbers and served at a temperature of 10 °C in acrylic cups, one at a time, in random order. The sensory panel consisted in 72 untrained judges who were habitual consumers of fruit juices. The project was approved by the Ethical Conduct in Research Involving Human Beings Committee from the Londrina State University, through opinion CEP/UEL 342/2011, n° CAAE 0319.0.268.000-11.

2.9. Experimental Design and Statistical Analysis

The physicochemical analysis and antioxidant activity were carried out according to a split plot design, in which the main treatment was the formulation and the secondary treatment was the storage duration. The experiment was replicated three times. In each replication, analyses were conducted in triplicate. The sensory experiment followed a randomized complete blocks design, where the juices were the treatments and the blocks the judges. The data were submitted to an analysis of variance (ANOVA) and the mean comparison Tukey test ($p < 0.05$). The statistical analyses were realized using the Statistical Analysis System version 9.1.3.

3. Results and Discussion

3.1. Physicochemical Characteristics

The results for pH and titratable acidity of the juices are presented in Table 1. The mixed formulations (BOMJ 1:1 and 1:2) presented titratable acidity (0.41%–0.59% citric acid) with intermediate values between beet juice (BJ or PBJ; 0.08%–0.14% citric acid) and orange juice (POJ; 0.77%–0.79% citric acid) ($p \le 0.05$), as expected. The same trend was observed for pH values, where the mixed formulations (BOMJ 1:1 and 1:2) presented pH (4–4.21) with intermediate values between beet juice (BJ or PBJ; 4.84–5.71) and orange juice (POJ; 3.83–3.88) ($p \le 0.05$). The presence of higher quantities of orange juice (BOMJ 1:2) in the mixed juice resulted in more acidic products, with a decrease in the pH values and an increase in the titratable acidity ($p \le 0.05$) when compared to the product with lower orange content (BOMJ 1:1). The pH and titratable acidities of the beet and orange juices are similar to those reported by other authors [18,19].

Organic acids contribute to the flavor and palatability of fruit juices. The greater acidity of the mixed juices (BOMJ 1:1 or BOMJ 1:2) when compared to the beet juices (BJ or PBJ) could protect them from the development of food spoilage micro-organisms, increasing their shelf life [19]. However, it can decrease the sensory acceptance of the products, because consumers are not attracted to very acidic products [20]. It is noteworthy, however, that the mixed juices (BOMJ 1:1 and BOMJ 1:2) still had lower acidity than the conventional orange juice (POJ).

Table 1. Titratable acidity and pH of juices stored at 4 °C for 30 days.

Juices	Titratable Acidity (% Citric Acid)				
	Storage Time (Days)				
	0	5	10	15	30
BJ	0.08 ± 0.00 cD	0.08 ± 0.00 cD	0.09 ± 0.00 cbD	0.09 ± 0.01 bD	0.14 ± 0.06 aA
PBJ	0.08 ± 0.00 aD	0.08 ± 0.00 aD	0.09 ± 0.00 aD	0.08 ± 0.00 aD	0.09 ± 0.01 aB
POJ	0.78 ± 0.00 abA	0.78 ± 0.01 abA	0.78 ± 0.00 abA	0.79 ± 0.00 aA	0.77 ± 0.01 bC
BOMJ 1:1	0.43 ± 0.01 aC	0.42 ± 0.00 bC	0.43 ± 0.01 aC	0.42 ± 0.00 bC	0.41 ± 0.00 bD
BOMJ 1:2	0.59 ± 0.01 aB	0.57 ± 0.01 bB	0.57 ± 0.00 bcB	0.56 ± 0.01 cB	0.54 ± 0.00 dE

Juices	pH				
	Storage Time (Days)				
	0	5	10	15	30
BJ	5.71 ± 0.02 aA	5.62 ± 0.01 abA	5.56 ± 0.02 bB	5.36 ± 0.21 cB	4.84 ± 0.26 dB
PBJ	5.70 ± 0.01 aA	5.69 ± 0.01 aA	5.68 ± 0.01 aA	5.68 ± 0.01 aA	5.49 ± 0.40 bA
POJ	3.88 ± 0.00 aD	3.83 ± 0.01 aD	3.87 ± 0.00 aE	3.83 ± 0.00 aE	3.88 ± 0.00 aE
BOMJ 1:1	4.17 ± 0.01 aB	4.18 ± 0.01 aB	4.16 ± 0.00 aC	4.21 ± 0.03 aC	4.16 ± 0.01 aC
BOMJ 1:2	4.03 ± 0.00 aC	4.00 ± 0.01 aC	4.02 ± 0.00 aD	4.03 ± 0.02 aD	4.03 ± 0.01 aD

The values represent the mean ± standard deviation ($n = 9$). Different superscript low-case letters in the same line indicate significant difference ($p < 0.05$) for each juice in relation to storage time. Different superscript upper-case letters in the same column indicate a significant difference ($p \leq 0.05$) between juices for the same storage day. BJ, natural beet juice; PBJ, pasteurized beet juice; POJ, pasteurized orange juice; BOMJ 1:1, pasteurized beet and orange mix juice 1:1 (v/v); BOMJ 1:2, pasteurized beet and orange mix juice 1:2 (v/v).

During the storage period, the titratable acidity increased ($p < 0.05$) in the BJ juice, decreased in the mixed juices (BOMJ 1:1 and BOMJ 1:2), and was maintained in the PBJ and POJ juices ($p > 0.05$). For the pH values, the products with beet (BJ and PBJ) exhibited a decrease ($p \leq 0.05$) in pH values during storage, and the products with added orange (POJ, BOMJ 1:1, and BOMJ 1:2) maintained ($p > 0.05$) this parameter. Probably, the presence of orange juice (POJ, BOMJ 1:1, and BOMJ 1:2) resulted in an increase in the buffering capacity of the juices, maintaining the pH of the products [20], which did not occur in the juices with beet only.

The color parameters (L*, a*, b* and Croma) are shown in Table 2. The total color difference is presented in Figure 1. The initial values of a* for BJ (0.78) and PBJ (0.95) indicate that the beet-added juices had a slightly red-purple color, probably originating from the presence of betacyanin. The b* values for BJ (1.56) and PBJ (1.67) represent a slight predominance of yellow by the presence of betaxanthins [21]. The POJ juice presented values of L* (36.61), a* (-2.49), and b* (16.34), demonstrating the yellow color of the orange juice. The mixed juices presented color characteristics of both juices (beet and orange) (L* = 23−24; a* = 4.59−8.76; and b* = 2.19−2.75).

During the storage period, all formulations had the same behavior for the L* and a* parameters, as the products stored for 30 days had higher L* and a* values than the newer ones (day 0) ($p \leq 0.05$). These results indicate that the juices became redder and with a lighter color. The juices (BJ, PBJ, POJ, BOMJ 1:1, and BOMJ 1:2) also had a decrease in the parameter b*, demonstrating a discoloration of the yellow color. In the POJ juice, a more intense yellow color was observed (i.e., an increase in b* values), and the total difference in color (ΔE^*ab) was the greatest (6.25). Therefore, the beet juice made the color of the juices more stable to the effects of refrigerated storage. The alterations in the color parameters were probably caused by the oxidative and non-oxidative reactions of polyphenols, resulting in colored condensation products. Furthermore, the Maillard Reaction or melanoidins' formation could cause the alteration [20].

Table 2. Color parameters for juices stored at 4 °C for 30 days

	L*				
Juices	**Storage Time (Days)**				
	0	**5**	**10**	**15**	**30**
BJ	22.45 ± 0.16 cD	22.87 ± 0.04 bD	22.31 ± 0.07 Cd	22.37 ± 0.02 cD	24.16 ± 0.47 aD
PBJ	22.37 ± 0.02 cD	22.71 ± 0.10 bD	22.34 ± 0.14 cD	22.23 ± 0.06 cD	23.72 ± 0.20 aE
POJ	36.61 ± 0.14 dA	38.18 ± 0.04 bA	36.73 ± 0.03 dA	37.04 ± 0.17 cA	42.42 ± 0.73 aA
BOMJ 1:1	23.21 ± 0.04 cC	23.76 ± 0.06 bC	23.15 ± 0.02 cC	23.14 ± 0.05 cC	24.78 ± 0.17 aC
BOMJ 1:2	23.77 ± 0.04 cB	24.33 ± 0.07 bB	23.81 ± 0.06 cB	24.15 ± 0.04 bB	26.18 ± 0.27 aB

	a*				
Juices	**Storage Time (Days)**				
	0	**5**	**10**	**15**	**30**
BJ	0.78 ± 0.06 bD	0.68 ± 0.03 cD	0.74 ± 0.03 bcD	0.81 ± 0.03 bC	1.07 ± 0.16 aD
PBJ	0.95 ± 0.03 bC	0.82 ± 0.04 cC	0.87 ± 0.06 bcC	0.91 ± 0.03 bcC	1.19 ± 0.13 aC
POJ	−2.49 ± 0.02 dE	−2.87 ± 0.02 bE	−2.51 ± 0.04 dE	−2.63 ± 0.05 cD	−3.64 ± 0.14 aE
BOMJ 1:1	4.80 ± 0.05 cB	5.18 ± 0.03 bB	4.78 ± 0.07 cB	4.59 ± 0.13 dB	6.13 ± 0.18 aB
BOMJ 1:2	6.65 ± 0.04 cA	7.07 ± 0.05 cA	6.65 ± 0.03 cA	6.68 ± 0.10 cA	8.76 ± 0.32 aA

	b*				
Juices	**Storage Time (Days)**				
	0	**5**	**10**	**15**	**30**
BJ	1.56 ± 0.04 aE	1.52 ± 0.03 aD	1.56 ± 0.03 aD	1.57 ± 0.03 aC	0.97 ± 0.16 bE
PBJ	1.67 ± 0.02 aD	1.61 ± 0.03 aD	1.63 ± 0.05 aD	1.64 ± 0.03 aC	1.21 ± 0.11 bD
POJ	16.34 ± 0.31 dA	17.03 ± 0.02 bA	16.46 ± 0.08 cA	15.95 ± 0.16 eA	18.34 ± 0.17 aA
BOMJ 1:1	2.39 ± 0.03 aC	2.38 ± 0.03 aC	2.35 ± 0.04 aC	2.19 ± 0.14 bB	2.23 ± 0.07 bC
BOMJ 1:2	2.75 ± 0.03 abB	2.82 ± 0.03 aB	2.68 ± 0.02 bB	2.26 ± 0.06 dB	2.47 ± 0.02 cB

	Chroma				
Juices	**Storage Time (Days)**				
	0	**5**	**10**	**15**	**30**
BJ	1.74 ± 0.03 aE	1.66 ± 0.02 aE	1.73 ± 0.03 aA	1.77 ± 0.02 aA	1.45 ± 0.12 bA
PBJ	1.92 ± 0.02 aD	1.80 ± 0.03 abD	1.85 ± 0.03 aB	1.88 ± 0.03 aA	1.69 ± 0.16 bB
POJ	16.53 ± 0.31 dA	17.27 ± 0.02 bA	16.65 ± 0.08 cC	16.16 ± 0.16 eB	18.70 ± 0.19 aC
BOMJ 1:1	5.36 ± 0.05 cC	5.70 ± 0.03 bC	5.33 ± 0.05 cD	5.08 ± 0.17 dC	6.52 ± 0.17 aD
BOMJ 1:2	7.19 ± 0.04 cB	7.61 ± 0.04 bB	7.17 ± 0.02 cdE	7.05 ± 0.11 dD	9.10 ± 0.31 aE

The values represent the mean ± standard deviation ($n = 9$). Different superscript low-case letters in the same line indicate significant difference ($p < 0.05$) for each juice in relation to storage time. Different superscript upper-case letters in the same column indicate significant difference ($p \leq 0.05$) between juices for the same storage day. BJ, raw beet juice; PBJ, pasteurized beet juice; POJ, pasteurized orange juice; BOMJ 1:1, pasteurized beet and orange mix juice 1:1 (v/v); BOMJ 1:2, pasteurized beet and orange mix juice 1:2 (v/v).

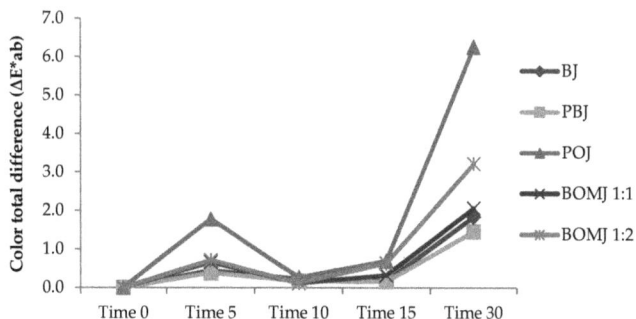

Figure 1. Total color difference (ΔE *ab) in juices stored at 4 °C for 30 days. BJ, raw beet juice; PBJ, pasteurized beet juice; POJ, pasteurized orange juice; BOMJ 1:1, pasteurized beet and orange mix juice 1:1 (v/v); BOMJ 1:2, pasteurized beet and orange mix juice 1:2 (v/v).

The physicochemical results demonstrated that the presence of orange juice in the mixed juices improved the pH stability of the products to the effects of refrigerated storage, as demonstrated by the maintenance of the pH of the orange and mixed juices. The alterations in the titratable acidity were slight (0.2% citric acid for BOMJ 1:1 and 0.05% citric acid for BOMJ 1:2). On the other hand, the presence of beet juice improved the color stability of the products, as the mixed juices showed a reduction of only 7%–10% in yellow color intensity and an increase in color purity (Chroma) by 22%–27%, probably due to the betalains' stability. These results reinforced the viability of the production of mixed juices with orange and beet. The physicochemical stability is desirable, because it confirms that the products remain similar to those that are newly manufactured, even after several weeks of storage [19].

3.2. Ascorbic Acid Content

The ascorbic acid content is shown in Figure 2. The initial values were 37.19, 25.93, and 39.84 mg/100 mL for POJ, BOMJ 1:1, and BOMJ 1:2, respectively, with the mixed juice with a higher quantity of orange juice (BOMJ 1:2) presenting higher ascorbic acid content than the other mixed juice (BOMJ 1:1), as expected. The pasteurized orange juice (POJ) showed an ascorbic acid loss of 25% after 30 days of storage ($p \leq 0.05$). The ascorbic acid stability is influenced by many factors, such as storage temperature, concentration of salt, sugars, and minerals, pH, oxygen levels, and the presence of enzymes or light [22]. However, the final values were in compliance with the legislation that established a minimum limit of ascorbic acid of 25 mg/100 mL for juice [23].

In the mixed juices (BOMJ 1:1 and BOMJ 1:2), a rapid loss of the ascorbic acid content occurred, with reduction right after 5 days of storage ($p \leq 0.05$). In BOMJ 1:1, the ascorbic acid content was completely lost at 15 days of storage, whereas in the BOMJ 1:2 juice a very low content was observed at day 30. The ascorbic acid can act as an antioxidant, removing the oxygen and controlling the activity of polyphenol oxidases enzymes, increasing the stability of the pigments [24]. In this study, it is possible that the decrease in the ascorbic acid content in the mixed juices is associated with a greater stability of the betalains pigments.

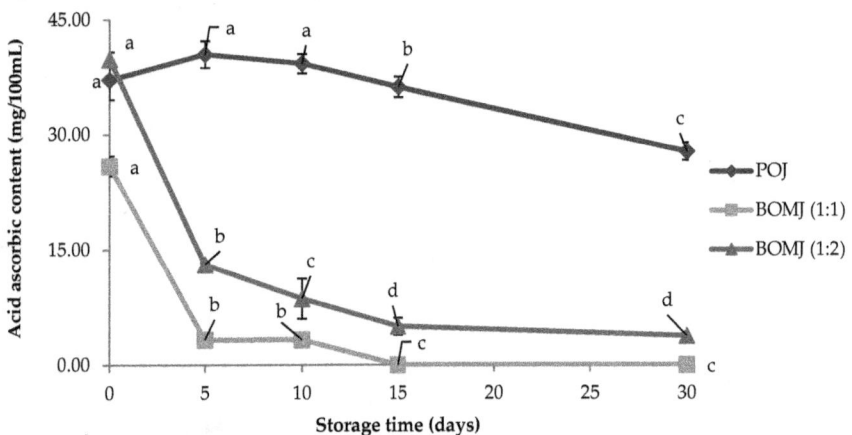

Figure 2. Acid Ascorbic Content (mg/100mL) in juices stored at 4 °C for 30 days. The values represent the mean ± standard deviation (*n* = 9). Different superscript lower case letters indicate significant difference (*p* < 0.05) for each juice in relation to storage time. POJ, pasteurized orange juice; BOMJ 1:1, pasteurized beet and orange mix juice 1:1 (*v/v*); BOMJ 1:2, pasteurized beet and orange mix juice 1:2 (*v/v*).

3.3. Betalain Content

Table 3 shows the results for the betacyanin and betaxanthin content during the refrigerated storage of the juices. The initial betacyanin content (Day 0) in pasteurized pure beet juice (PBJ) (53.39 mg/100 mL) was lower than the content in the raw beet juice (BJ) (65.32 mg/100 mg), showing that the pasteurization caused a reduction in betacyanins by 18% ($p \leq 0.05$). For the betaxanthins, the thermal treatment had no significant effect ($p > 0.05$).

Table 3. Betacyanin and betaxanthins content (mg/100 mL) in juices stored at 4 °C for 30 days.

Juices	Storage Time (Days)				
	0	5	10	15	30
Betacyanin					
BJ	65.32 ± 1.58 aA	61.18 ± 1.72 bA	55.34 ± 2.09 cA	51.76 ± 3.51 cA	40.64 ± 7.79 dA
PBJ	53.39 ± 3.31 aB	50.52 ± 5.89 bcB	49.48 ± 2.61 cB	42.31 ± 3.68 dB	31.55 ± 5.64 eB
BOMJ 1:1	31.12 ± 0.89 aC	31.99 ± 0.59 aC	31.63 ± 0.76 aC	28.27 ± 5.19 bC	30.29 ± 1.94 abC
BOMJ 1:2	22.31 ± 0.98 abD	23.17 ± 0.77 aD	22.85 ± 0.78 abD	20.37 ± 3.77 bD	22.32 ± 1.40 abD
Betaxanthins					
BJ	29.99 ± 1.86 aA	28.84 ± 0.79 abB	28.42 ± 2.34 bB	27.82 ± 2.77 bA	22.79 ± 5.61 cB
PBJ	31.76 ± 0.51 aA	30.43 ± 0.08 bA	30.40 ± 2.29 bA	27.80 ± 1.44 cA	25.69 ± 0.22 dA
BOMJ 1:1	11.96 ± 0.30 abB	12.08 ± 0.16 aC	11.76 ± 0.25 abC	11.44 ± 0.29 abB	10.74 ± 0.93 bC
BOMJ 1:2	8.80 ± 0.29 aC	8.84 ± 0.16 aD	8.52 ± 0.16 aD	8.21 ± 0.18 aC	7.82 ± 0.60 aD

The values represent the mean ± standard deviation ($n = 9$). Different superscript low-case letters in the same line indicate significant difference ($p < 0.05$) for each juice in relation to storage time. Different superscript upper-case letters in the same column indicate significant difference ($p \leq 0.05$) between juices for the same storage day, BJ (raw beet juice), PBJ (pasteurized beet juice), BOMJ 1:1 (pasteurized beet and orange mix juice 1:1 (v/v) and BOMJ 1:2 (pasteurized beet and orange mix juice 1:2 (v/v)).

The betacyanin and betaxanthins content of the mixed juices BOMJ 1:1 (31.12 mg/100 mL and 11.96 mg/100 mL) and BOMJ 1:2 (22.31 mg/100 mL e 8.80 mg/100 mL) were lower ($p \leq 0.05$) than those of the juices only with beet (BJ with 65.32 mg/100 mL and 29.99 mg/100 mL and PBJ with 53.39 mg/100 mL e 31.76 mg/100 mL). These results were expected, due to the dilution of the beet juice with the orange juice, proportional to each mixed juice formulation.

At the end of the storage time, the beet juices (BJ and PBJ) showed a significant reduction of 37–41% in betacyanin content, while for the betaxanthins, a reduction of 19–24% was observed. In the mixed juices (BOMJ 1:1 and BOMJ 1:2), the betacyanin (22.31–31.63 mg/100 mL) and betaxanthin (7.82–12.08 mg/100 mL) content was maintained, when comparing the products at 0 and 30 days of storage ($p > 0.05$). As discussed before, the results demonstrate that the orange juice contributed to the stability of both classes of betalains in the mixed juices, probably due to the action of the citric acid and the ascorbic acid. The citric acid is an antioxidant and chelating agent, and acts synergistically with the ascorbic acid. The maintenance of these beneficial compounds in the mixed juices is of primary importance in the development of functional foods.

3.4. Total Phenolic Compounds Content and Antioxidant Activity

Total phenolic content and the antioxidant activity (DPPH Trolox or ABTS trolox) are presented in Table 4. All juices were significant sources of polyphenols. The raw beet juice (BJ) and the pasteurized beet juice (PBJ) presented 521 and 497 µg EAG/mL, respectively; the mixed juices had 484–485 µg EAG/mL; and the orange juice (POJ) had 448 µg EAG/mL at day 0. In fact, orange juices are good sources of polyphenolic compounds, including hydroxycinnamic acids and flavonoids (mainly flavonas) [25], while beet juice has a high antioxidant action and is a rich source of polyphenols. Among the beet extract phenolic acids, the 4-hydroxybenzoic acid is the main constituent, followed by the cynamic, vanillic, and chlorogenic acids, the trans ferulic acid, and caffeic acid [2].

During storage time, the beet juices (BJ and PBJ) showed a significant loss ($p \leq 0.05$) of phenolic compounds by 14% and 13%, respectively, while the orange and mixed juices did not show any significant variation ($p > 0.05$) when comparing the products at day 0 and 30 of storage. This was probably due to the joint action of the citric and ascorbic acids, which inhibited the oxidative reactions of the phenolic compounds presented in the juices, demonstrating the beneficial effect of the addition of orange juice to the beet juice.

In both methodologies of antioxidant activity evaluation (DPPH and ABTS methods), the samples with greater antioxidant activity were the raw (2733 and 2179 μg Trolox/mL) and pasteurized beet (2813 and 2081 μg Trolox/mL) juices, and the sample with the lowest antioxidant activity was the orange juice (1408 and 1872 μg Trolox/mL). In fact, betalains' antioxidant activity is greater than that of ascorbic acid, and beet is one of the most powerful green vegetables with regard to antioxidant activity [26], when beet juice is compared to other fruits that are well-accepted as having high antioxidant contents, such as pomegranate and cranberry [27]. The mixed juices had intermediated antioxidant activity, with 1930–2083 μg Trolox/mL in the DPPH method and 1840–1854 μg Trolox/mL in the ABTS method.

During storage, the antioxidant activity decreased in all of the juices evaluated in one or both methodologies (DPPH or ABTS) ($p \leq 0.05$). Pasteurized beet juice (PBJ) showed the greatest loss (37%) in the DPPH method, followed by POJ (32%), BJ (21%), BOMJ 1:2 (19%), and BOMJ 1:1 (no significant loss). In the ABTS methodology, the decrease was: POJ (21%), PBJ (14%), BOMJ 1:1 (13%), and BOMJ 1:2 (12%). The decrease in ascorbic acid and total phenolic compounds is the probable cause for the reduction in antioxidant capacity in the juice samples analyzed in this study. Therefore, this study demonstrated that the orange juice maintained the stability of the phenolic compounds, and reduced the loss of the antioxidant activity of the juices.

Table 4. Total Phenolic compounds (μg EAG/mL), DPPH Trolox (μg/mL), and ABTS Trolox (μg/mL) in juices stored at 4 °C/30 days.

Juices	Total Phenolic Compounds				
	Storage Time (Days)				
	0	5	10	15	30
BJ	521 ± 19 bA	518 ± 9 bA	571 ± 86 aA	515 ± 62 bA	448 ± 48 cB
PBJ	497 ± 26 bABD	494 ± 20 bAB	555 ± 76 aA	517 ± 54 bA	431 ± 19 cB
POJ	448 ± 19 abC	473 ± 10 aB	477 ± 18 aBD	458 ± 15 abB	433 ± 41 bB
BOMJ 1:1	484 ± 33 cDB	499 ± 18 abcAB	519 ± 50 aC	514 ± 48 abA	485 ± 65 bcA
BOMJ 1:2	485 ± 50 abB	506 ± 29 aA	505 ± 42 aCD	490 ± 42 aA	459 ± 59 bAB
DPPH					
BJ	2733 ± 188 aA	2527 ± 255 abA	2307 ± 325 bcA	2323 ± 419 bA	2157 ± 243 cA
PBJ	2813 ± 264 aA	2210 ± 285 bB	2040 ± 296 bcB	1870 ± 236 cdB	1760 ± 261 dB
POJ	1408 ± 245 aC	773 ± 96 bcD	1037 ± 153 bD	797 ± 167 bcD	957 ± 135 bcC
BOMJ 1:1	2083 ± 89 aB	2027 ± 187 abBC	1317 ± 318 cC	1670 ± 369 bBC	1853 ± 231 abB
BOMJ 1:2	1930 ± 246 aB	1803 ± 146 abC	1452 ± 120 cC	1596 ± 200 bcC	1554 ± 119 cB
ABTS					
BJ	2179 ± 126 aA	1972 ± 209 bA	1950 ± 173 bA	2059 ± 145 bA	1720 ± 352 cAD
PBJ	2081 ± 142 aA	1741 ± 155 cB	1957 ± 206 bA	1999 ± 63 abA	1798 ± 62 cA
POJ	1872 ± 241 aB	1515 ± 93 bC	1507 ± 103 bcB	1575 ± 148 bB	1400 ± 113 cB
BOMJ 1:1	1854 ± 75 aB	1746 ± 136 abB	1642 ± 61 bcC	1711 ± 92 bcC	1604 ± 54 cC
BOMJ 1:2	1840 ± 131 aB	1667 ± 68 bB	1552 ± 51 cBC	1605 ± 84 bcBC	1614 ± 146 bcCD

The values represent the mean ± standard deviation ($n = 9$). Different superscript low-case letters in the same line indicate significant difference ($p < 0.05$) for each juice in relation to storage time. Different superscript upper-case letters in the same column indicate significant difference ($p \leq 0.05$) between juices for the same storage day. BJ, raw beet juice; PBJ, pasteurized beet juice; POJ, pasteurized orange juice; BOMJ 1:1, pasteurized beet and orange mix juice 1:1 (v/v); BOMJ 1:2, pasteurized beet and orange mix juice 1:2 (v/v).

3.5. Sensory Acceptance

Consumer acceptance is presented in Table 5. Pasteurized beet juice (PBJ) acceptance was lower ($p \leq 0.05$) than that of the mixed juices (BOMJ 1:1 and BOMJ 1:2) with regard to the attributes aroma, flavor, and overall acceptance. For the color attribute, it was not different from the BOMJ 1:2 ($p > 0.05$). The PBJ juice had satisfactory acceptance in the color parameter (a score of 7.5 in a 9-point hedonic scale), but, in aroma, flavor, and overall acceptance, the scores were low (<5.9).

Table 5. Acceptance of the juices.

Juices	Color	Aroma	Flavor	Overall Acceptance
PBJ	7.5 ± 1.8 b	5.9 ± 1.7 b	4.9 ± 2.2 b	5.6 ± 1.8 b
BOMJ (1:1)	8.0 ± 0.9 a	6.6 ± 1.7 a	6.3 ± 1.7 a	6.8 ± 1.5 a
BOMJ (1:2)	7.8 ± 1.1 ab	6.9 ± 1.9 a	6.6 ± 2.0 a	6.9 ± 1.8 a

The values represent the mean ± standard deviation, where different lower case superscript letters in the same column indicate significant differences ($p < 0.05$) between the juices for the same attribute or overall acceptance. PBJ, pasteurized beet juice; BOMJ 1:1, pasteurized beet and orange mix juice 1:1 (v/v); BOMJ 1:2, pasteurized beet and orange mix juice 1:2 (v/v). On the scale of acceptance, 1 = extremely dislike and 9 = like extremely.

The addition of orange juice to the beet juice increased the acceptance of the products in all evaluated parameters (color, aroma, flavor, and overall acceptance). The mixed juices received scores between 6.3 and 8 in a 9-point hedonic scale, indicating that the consumers liked the products from slightly to much. The color was the attribute with the highest score (7.8–8), probably because of the intensity and attractiveness that the beet gives to the product due to the presence of betalains.

The results of the present study demonstrate the applicability of the mixed juices, as the products were well accepted. The pasteurized beet juice (PBJ) had a higher content of betacyanin, betaxanthin, total phenolic content, and antioxidant activity; however, it was not well accepted by consumers, and the functional components were less stable to the effects of refrigerated storage in this juice.

Currently, several studies have focused on the antioxidant capacity and consumption benefits of the fruit and green vegetables juices, including beet juice [2,28,29]; however, there are no data related to the sensory acceptance of these types of beverages by population.

4. Conclusions

The beet and orange mixed juices developed in this study could provide consumers with strong natural antioxidant compounds and bioactive phytochemicals, and they had good acceptance. The presence of orange juice contributed to the pH, betacyanin, betaxanthin, and antioxidant capacity stabilities during storage, whereas the presence of beet improved the color stability. The mixed juices showed high total phenolic compounds (484–485 µg gallic acid/mL), DPPH scavenging capacity (2083–1930 µg Trolox/mL), and ABTS (1854–1840 µg Trolox/mL). The beet and orange mixed juice is an alternative functional beverage that can contribute to an increase in the consumption of beet and orange.

Acknowledgments: The authors gratefully acknowledge the financial support of Conselho Nacional de Pesquisa (CNPq) and Fundação Araucária.

Author Contributions: All authors contributed equally to this work.

Conflicts of Interest: The authors declare no conflict of interest.

References

1. FAO. Crop Water Information: Sugarbeet. 2015. Available online: http://www.fao.org/nr/water/cropinfo_sugarbeet.html (accessed on 22 February 2017).
2. Wootton-Beard, P.C.; Moran, A.; Ryan, L. Stability of the antioxidant capacity and total polyphenol content of 23 commercially available vegetable jucies before and after in vitro digestion as measured by FRAP, DPPH, ABTS and Folin Ciocalteau methods. *Food Res. Int.* **2011**, *44*, 217–224. [CrossRef]
3. Wruss, J.; Waldenberger, G.; Huemer, S.; Uygun, P.; Lazerstorfer, P.; Müller, U.; Höglinger, O.; Weghuber, J. Compositional characteristics of commercial beetroot products and beetroot juice prepared from seven beetroot varieties grown in Upper Austria. *J. Food Comps. Anal.* **2015**, *42*, 46–55. [CrossRef]
4. Gengatharan, A.; Dykes, G.A.; Choo, W.S. Betalains: Natural plant pigments with potential application in functional foods. *LWT Food Sci. Technol.* **2015**, *64*, 645–649. [CrossRef]

5. Farabegoli, F.; Scarpa, E.S.; Frati, A.; Serafini, G.; Papi, A.; Spisni, E.; Antonini, E.; Benedetti, S.; Ninfali, P. Betalains increase vitexin-2-O-xyloside cytotoxicity in CaCo-2 cancer cells. *Food Chem.* **2017**, *218*, 356–364. [CrossRef] [PubMed]

6. Tarrago-Trani, M.T.; Phillips, K.M.; Cotty, M. Matrix-specific method validation for quantitative analysis of vitamin c in diverse foods. *J. Food Comps. Anal.* **2012**, *26*, 12–25. [CrossRef]

7. Oikeh, E.I.; Omoregie, E.S.; Oviasogie, F.E.; Oriakhi, K. Phytochemical, antimicrobial, and antioxidant activities of different citrus juice concentrates. *Food Sci. Nutr.* **2016**, *4*, 103–109. [CrossRef] [PubMed]

8. Fusco, R.; Cirmi, S.; Gugliandolo, E.; di Paola, R.; Cuzzocrea, S.; Navarra, M. A flavonoid-rich extract of orange juice reduced oxidative stress in an experimental model of inflammatory bowel disease. *J. Funct. Foods* **2017**, *30*, 168–178. [CrossRef]

9. FAO. Food Based Dietary Guidelines for Sri Lankans. 2011. Available online: http://www.fao.org/nutrition/ education/food-based-dietary-guidelines/regions/countries/sri-lanka/en/ (accessed on 22 February 2017).

10. Vigitel Brasil 2011: Vigilância de Fatores de Risco e Proteção para Doenças Crônicas por Inquérito Telefônico. Available online: http://bvsms.saude.gov.br/bvs/publicacoes/vigitel_brasil_2011_fatores_risco_doencas_ cronicas.pdf (accessed on 18 July 2017).

11. Wootton-Beard, P.C.; Ryan, L. Improving public health?: The role of antioxidant-rich fruit and vegetable beverages. *Food Res. Int.* **2011**, *44*, 3135–3148. [CrossRef]

12. Association of Official Analytical Chemistry (AOAC). *Official Methods of Analysis*, 15th ed.; Association of Official Analytical Chemists: Washington, DC, USA, 2004.

13. Souza, M.C.C.; Benassi, M.T.; Meneghel, R.F.A.; Silva, R.S.S.F. Stability of unpasteurized and refrigerated orange juice. *Braz. Arch. Biol. Technol.* **2004**, *47*, 391–397. [CrossRef]

14. Stintzing, F.C.; Schieber, A.; Carle, R. Evaluation of colour properties and chemical quality parameters of cactus juices. *Eur. Food Res. Technol.* **2003**, *216*, 303–311. [CrossRef]

15. Singleton, V.L.; Orthofer, R.; Lamuela-Raventos, R.M. Analysis of total phenols and other oxidation substrates and antioxidants by means of Folin–Ciocalteu reagent. *Methods Enzymol.* **1999**, *299*, 152–178.

16. Brand-Williams, W.; Cuvelier, M.E.; Berset, C. Use of a free-radical method to evaluate antioxidant activity. *LWT Food Sci. Technol.* **1995**, *28*, 25–30. [CrossRef]

17. Ozgen, M.; Reese, R.N.; Tulio, A.Z.; Scheerens, J.C.; Miller, A.R. Modified 2,2- azino-bis-3-ethylbenzothiazoline-6-sulfonic acid (ABTS) method to measure antioxidant capacity of selected small fruits and comparison to ferric reducing antioxidant power (FRAP) and 2,2-diphenyl-1-picrylhydrazyl (DPPH) methods. *J. Agric. Food Chem.* **2006**, *54*, 1151–1157. [CrossRef] [PubMed]

18. Yoon, K.Y.; Woodams, E.E.; Hang, Y.D. Fermentation of beet juice by beneficial lactic acid bacteria. *LWT Food Sci. Technol.* **2005**, *38*, 73–75. [CrossRef]

19. Pimentel, T.C.; Madrona, G.S.; Garcia, S.; Prudencio, S.H. Probiotic viability, physicochemical characteristics and acceptability during refrigerated storage of clarified apple juice supplemented with *Lactobacillus paracasei* ssp. *paracasei* and oligofructose in different package type. *LWT Food Sci. Technol.* **2015**, *63*, 415–422. [CrossRef]

20. Costa, G.M.; Silva, J.V.C.; Mingotti, J.D.; Barão, C.E.; Klososki, S.J.; Pimentel, T.C. Effect of ascorbic acid or oligofructose supplementation on L. paracasei viability, physicochemical characteristics and acceptance of probiotic orange juice. *LWT Food Sci. Technol.* **2017**, *75*, 195–201. [CrossRef]

21. Delgado-Vargas, F.; Jiménez, A.R.; Paredes-López, O. Natural pigments: Carotenoids, anthocyanins, and betalains—Characteristics, biosynthesis, processing, and stability. *Crit. Rev. Food Sci. Nutr.* **2000**, *40*, 173–289. [CrossRef] [PubMed]

22. Tiwari, B.K.; Donnell, C.P.O.; Muthukumarappan, K.; Cullen, P.J. Ascorbic acid degradation kinetics of sonicated orange juice during storage and comparison with thermally pasteurised juice. *LWT Food Sci. Technol.* **2009**, *42*, 700–704. [CrossRef]

23. *Ministério da Agricultura e do Abastecimento Instrução Normativa n° 1, de 7 de Janeiro de 2000. Complementa Padrões de Identidade e Qualidade Para Suco de Laranja*; Diário Oficial da União da República Federativa do Brasil: Brasília (DF), Brasil, 2000; Seção 1; p. 54.

24. Cardoso-Ugarte, G.A.; Sosa-Morales, M.E.; Ballard, T.; Liceaga, A.; San Martín-González, M.F. Microwave-assisted extraction of betalains from red beet (*Beta vulgaris*). *LWT Food Sci. Technol.* **2014**, *59*, 276–282. [CrossRef]

25. Gliszczynska-Swiglo, A.; Wroblewska, J.; Lemanska, K.; Klimczak, I.; Tyrakowska, B. The contribution of polyphenols and vitamin C to the antioxidant activity of commercial orange juices and drinks. In Proceedings of the 14th IGWT Symposium Focussing New Century, Commodity-trade Environment, Beijng, China, August 2004; pp. 121–126.

26. Zitnanova, I.; Ranostajova, S.; Sobotova, H.; Demelova, D.; Pechan, I.; Durackova, Z. Antioxidative activity of selected fruits and vegetables. *Biologia* **2006**, *61*, 279–284. [CrossRef]

27. Ryan, L.; Prescott, S.L. Stability of the antioxidant capacity of twenty-five commercially available fruit juices subjected to an in vitro digestion. *Int. J. Food Sci. Technol.* **2010**, *45*, 1191–1197. [CrossRef]

28. Ravichandran, K.; Ahmed, A.R.; Knorr, D.; Smetanska, I. The effect of different processing methods on phenolic acid content and antioxidant activity of red beet. *Food Res. Int.* **2012**, *48*, 16–20. [CrossRef]

29. Ravichandran, K.; Saw, N.M.M.T.; Mohdaly, A.A.A.; Gabr, A.M.M.; Kastell, A.; Riedel, H.; Cai, Z.; Knorr, D.; Smetanska, I. Impact of processing of red beet on betalain content and antioxidant activity. *Food Res. Int.* **2013**, *50*, 670–675. [CrossRef]

Article

LC–MS/MS and UPLC–UV Evaluation of Anthocyanins and Anthocyanidins during Rabbiteye Blueberry Juice Processing

Rebecca E. Stein-Chisholm [1], John C. Beaulieu [2,*] ⓘ, Casey C. Grimm [2] and Steven W. Lloyd [2]

[1] Lipotec USA, Inc. 1097 Yates Street, Lewisville, TX 75057, USA; rstein07@gmail.com
[2] United States Department of Agriculture, Agricultural Research Service, Southern Regional Research Center, 1100 Robert E. Lee Blvd., New Orleans, LA 70124, USA; Casey.Grimm@ars.usda.gov (C.C.G.); Steven.Lloyd@ars.usda.gov (S.W.L.)

Academic Editor: António Manuel Jordão
Received: 11 October 2017; Accepted: 1 November 2017; Published: 25 November 2017

Abstract: Blueberry juice processing includes multiple steps and each one affects the chemical composition of the berries, including thermal degradation of anthocyanins. Not-from-concentrate juice was made by heating and enzyme processing blueberries before pressing, followed by ultrafiltration and pasteurization. Using LC–MS/MS, major and minor anthocyanins were identified and semi-quantified at various steps through the process. Ten anthocyanins were identified, including 5 arabinoside and 5 pyrannoside anthocyanins. Three minor anthocyanins were also identified, which apparently have not been previously reported in rabbiteye blueberries. These were delphinidin-3-(p-coumaroyl-glucoside), cyanidin-3-(p-coumaroyl-glucoside), and petunidin-3-(p-coumaroyl-glucoside). Delphinidin-3-(p-coumaroyl-glucoside) significantly increased 50% after pressing. The five known anthocyanidins—cyanidin, delphinidin, malvidin, peonidin, and petunidin—were also quantitated using UPLC–UV. Raw berries and press cake contained the highest anthocyanidin contents and contribute to the value and interest of press cake for use in other food and non-food products. Losses of 75.7% after pressing and 12% after pasteurization were determined for anthocyanidins during not-from-concentrate juice processing.

Keywords: not-from-concentrate juice; *Vaccinium ashei*; juice processing; anthocyanins; anthocyanidins

1. Introduction

Blueberries are a well-known source of health-promoting phytochemicals [1,2]. These phytochemicals can be divided into different classes based on their chemical structures. Of the major classes of phytochemicals, flavonoids are becoming popular for studies focusing on their health benefits. One of the most unique classes of flavonoids are the anthocyanins. Anthocyanins give fruits and vegetables their vibrant red, blue, and purple colors [3]. These compounds are unique because they exist in five configurations and various colors based on pH. These configurations include the blue-colored anionic quinonoidal base, the violet-colored quinonoidal base, the red-colored flavylium cation, the colorless carbinol base, and the yellow-colored (E) or (Z) chalcone [4]. At lower pH values, anthocyanins are red in color; as pH rises anthocyanins will be more blue in color but, as the pH shifts from more neutral to basic, they turn clear, then above pH 10, the alkalinity generally destroys the compound [5].

There have been over 600 naturally occurring anthocyanins reported in plants [6]. Anthocyanins are composed of an anthocyanidin backbone with varying glycosides. The five major anthocyanidin classes in blueberries are cyanidin, delphinidin, malvidin, peonidin and petunidin [7]. Cyanidin is the most common anthocyanidin found in plants and can have 76 different glycoside combinations [8]. The major anthocyanins in blueberries include 3-glycosidic derivatives of cyanidin, delphinidin, malvidin, peonidin and petunidin; with glucose, galactose and arabinose as the most abundant sugars [9]. The lesser anthocyanins consist of acetoly, malonoyl, and coumaroyl conjugated compounds [10].

Consumer demand for food and beverage products which are made from locally or regionally grown raw materials are driven by the belief that these products help local communities, provide healthier alternatives, as well as decrease carbon footprints [11]. With increased demand for natural and less processed food options, local small scale juice producers are looking at not-from-concentrate (NFC) juice products to meet demand and create niche markets [12]. In Europe, the NFC juice market segment was up 5.4% in 2016 from 2015 [13]. In the United States, NFC juices have an expected annual average growth of 5.3% through 2016 [14]. Utilizing NFC juices as a compromise between unpasteurized fresh juices and highly processed juices reconstituted from concentrates allows small-scale local producers to expand beyond the farmers markets [15,16]. Furthermore, the ability to process berries, especially locally frozen fruit into juice is an efficient way to extend their shelf life and extend the profitability of a grower's harvest season [17].

The process of making berry juice may include heating before pressing as well as enzyme treatments (creating a mash) to increase juice recovery and minimize anthocyanin loss [18]. Processing can affect the anthocyanins and other phytochemicals, as well as the macronutrients in the berries [19]. It is proposed that one of the pathways of degradation of anthocyanins is caused by native enzymes, mainly polyphenol oxidase (PPO), breaking down other polyphenols to form quinones. These in turn react with the anthocyanins, forming brown pigments [19]. Heating the berries before pressing denatures native enzymes and reduces enzymatic browning in juice [20]. The use of pectinases increases juice recovery by degrading pectin in cell walls, improving liquefaction and clarification, and aiding in filtration processes [21]. Heated mash is then pressed, removing remaining skins and seeds, resulting in an unfiltered juice. Filtration is an optional step to reduce sedimentation to clarify the juice and remove polymeric compounds which can affect overall color and turbidity [22]. Pasteurization is a safety step used to decrease spoilage and contamination. Evaluating the aforementioned processes builds information and awareness to help juice producers develop juice products that can maximize profit and quality.

The "Tifblue" variety at one point was the most widely planted rabbiteye (RAB) blueberry in the world [23]. It is still a very popular RAB berry cultivar today and is highly regarded for its appearance, productivity, harvesting and shipping qualities, as well as a standard for comparison to other selections and cultivars [24,25]. Many studies have been conducted on the processing effects on blueberry juice anthocyanins, but to our knowledge, few have been carried out on RAB (*Vaccinium ashei*) blueberries [26,27]. In this experiment, RAB blueberry anthocyanins and their anthocyanidin backbones were identified in each juice processing step and their stepwise changes were evaluated. This NFC juice evaluation contributes to the knowledge of RAB blueberry properties and pilot plant process parameters affecting polyphenolics during various juice processing steps.

2. Materials and Methods

2.1. Juice Processing

Commercial "Tifblue" RAB blueberries (*V. ashei*) were harvested (Blue River Farms, Hattiesburg, MS, USA) and commercially packaged (sorted, graded, washed, air-dried and forced-air rapid frozen at -20 °C (Nordic Cold Storage, Hattiesburg, MS, USA). Using a 37.9 L steam-jacketed kettle (Groen-A Dover Industries Co., Byram, MS, USA) 27 kg (two individual 13.5 kg boxes) of frozen berries (control) were quickly heated to ~95 °C in roughly 14.5 min with constant stirring using a large wooden paddle. Temperature was monitored with thermal probes ("K Milkshake"; ThermoWorks, Salt Lake City, UT, USA) and the crude mash was held at 96.5 ± 1.1 °C for 3 min. The mash (sampled prior to enzyme treatment and hydraulic pressing) was then poured into a 37.9 L stainless steel vessel and allowed to cool to 55 °C for addition of pectinase enzyme. Rohapect 10 L (AB Enzymes, Darmstadt, Germany) was added at 200 mL ton^{-1} and allowed to activate with occasional stirring for 1 h. The enzyme-treated mash was pressed warm (~45 °C) in an X-1 single-layer hydraulic press (Goodnature, Orchard Park, NY, USA) at 12.4 MPa using a medium-weave polyester mesh press bag (Goodnature, #2636) for 1 min. Pressed juice from each batch was individually collected in a stainless-steel vessel and cooled overnight at 4 °C. Press cake (PRC) samples were collected and stored at −20 °C.

Half of the chilled pressed juice (PJ) was portioned off and pasteurized, delivering a pasteurized pressed unfiltered juice (PPJ) and samples were collected at this point and stored at −20 °C. The remaining pressed juice was filtered using ultrafiltration in a pilot unit (BRO/BUF, Membrane Specialists, Hamilton, OH, USA) with a 100 L hopper tank. The unit consisted of an in-line membrane filtration module (PCI B-1 Module Series, Aquious PCI Membrane, Hamilton, OH, USA) and a heat exchanger fed by a 7.5 hp screw pump. Filtration occurred with a 200,000 molecular weight cut-off (0.2 μm) XP-201 polyvinylidene fluoride (PVDF) membrane (ITT PCI Membrane Systems, Zelienople, PA, USA), with the heat exchanger run at ambient (~25 °C) attaining a product flow rate of roughly 18.9–29.9 L h^{-1}. After equilibrating the ultrafiltration unit, filtered not-from-concentrate blueberry juice was collected for sampling (UF; ultrafiltered juice), then pasteurized (UFP, ultrafiltered pasteurized).

PJ and UF samples were pasteurized using a high-temperature short-time (HTST) pasteurization unit (Electra UHT/HTST Lab-25EDH; MicroThermics, Raleigh, NC, USA) at 90 °C for 10 s, at 1.2 L min^{-1}, followed by hot-filling at 85 °C into pre-sterilized 250 mL transparent glass media bottles (Corning, Tewksbury, MA, USA) followed by inversion and ice water bath chilling. Pressed juice which was pasteurized (PPJ) and ultrafiltered pasteurized juice (UFP) were frozen at −20 °C before anthocyanins and anthocyanidins analysis.

2.2. Anthocyanin Analysis

2.2.1. Extraction

For the control and PRC samples, 5 g of raw berries were thawed and homogenized using a Tekmar Tissumizer (SDK-1810, IKA-Werke, Staufen, Germany). Using 2 g of sample, all triplicated samples were lyophilized in a VirTis Genesis 25ES freeze dryer (SP Scientific, Warminster, PA, USA). After lyophilization, 100 mg of the powder was weighed into 2 mL centrifuge tubes and 1 mL of an extraction solvent (70:30:1, $v/v/v$; methanol (MeOH): water (H_2O): trifluoroacetic acid (TFA)) was added [10]. The tubes were vortexed for 15 s and left undisturbed for 60 min. Following extraction, the tubes were sonicated for 20 min and centrifuged (IEC CL, International Equipment Company, Needham Heights, MA, USA) for 15 min at 1200 rpm. The supernatant was filtered through a 0.2 μm syringe filter into a HPLC vial then stored at −20 °C.

2.2.2. HPLC–MS/MS Chromatography

Berry and juice samples were analyzed for anthocyanins using a LC–MS/MS method. Extracted samples were analyzed on an Agilent 1200 HPLC with an Agilent Small Molecule Chip Cube interface and Agilent 6520 Q-TOF MS/MS (Agilent, Santa Clara, CA, USA). The chip contained a 40 μL enrichment column and a C18 (43 mm × 75 μL, 80 Å) column. The eluents were acidified H_2O with 0.1% formic acid (A) and 90% acetonitrile with 9.9% H_2O and 0.1% formic acid (B). The gradient was held at 2% B then raised to 20% over 10 min, and then increased to 40% B to 18 min. The MS fragmenter was set to 175 V, and the VCap at 1800 V. Capillary temperature was 300 °C with N_2 as carrier gas with a flow rate of 5 L min^{-1}. The MS scan rate was 1 scan s^{-1} (10,000 transients). Auto MS/MS had selected m/z ranges (Table 1) and the scan rate was 1 scan s^{-1}. A semi-quantified peak area abundance was calculated for each compound using the characteristic anthocyanin parent fragment molecular weight ([M + H]$^+$) along with the fragmented MS/MS backbone anthocyanidin molecular weight (Table 1), as confirmed by the residual sugar (generally the 3 position in the C-ring or R3; the R-O-sugar group) moiety fragment. Averaged ion counts from these identifying fragments were utilized to measure and compare process changes in the anthocyanins in juices.

Table 1. Major and minor anthocyanin compounds and mass spectrometry variables found in blueberries.

Major Anthocyanin [z]	Sugar Moiety	Molecular Formula	[M + H]+ (m/z) [y]	MS/MS (m/z) [x]	R_t (min) HPLC [w]
Delphinidin-3-arabinoside	Arabinose	$C_{20}H_{19}O_{11}$	435.0922	303.0500	10.94
Cyanidin-3-arabinoside	Arabinose	$C_{20}H_{19}O_{10}$	419.0928	287.0550	11.47
Petunidin-3-arabinoside	Arabinose	$C_{21}H_{21}O_{11}$	449.1078	317.0700	11.66
Peonidin-3-arabinoside	Arabinose	$C_{21}H_{21}O_{11}$	433.1129	301.0700	12.20
Malvidin-3-arabinoside	Arabinose	$C_{22}H_{23}O_{11}$	463.1235	331.0800	12.34
Delphinidin-3-pyranoside	Galactose/Glucose	$C_{21}H_{21}O_{12}$	465.1027	303.0500	10.38
Cyanidin-3-pyranoside	Galactose/Glucose	$C_{21}H_{21}O_{11}$	449.1078	287.0550	10.99
Petunidin-3-pyranoside	Galactose/Glucose	$C_{22}H_{23}O_{12}$	479.1184	317.0700	11.29
Peonidin-3-pyranoside	Galactose/Glucose	$C_{22}H_{23}O_{11}$	463.1235	301.0700	12.99
Malvidin-3-pyranoside	Galactose/Glucose	$C_{23}H_{25}O_{12}$	493.1340	331.0800	11.98
Minor Anthocyanin [z]	**Sugar Moiety**	**Molecular Formula**	**[M + H]+ (m/z)**	**MS/MS (m/z)**	**R_t (min) HPLC [w]**
Delphinidin-3-(p-coumaroyl-glucoside)	Glucose	$C_{30}H_{27}O_{14}$	611.1395	303.0500	13.11
Cyanidin-3-(p-coumaroyl-glucoside)	Glucose	$C_{30}H_{27}O_{13}$	595.1446	287.0550	13.86
Petunidin-3-(p-coumaroyl-glucoside)	Glucose	$C_{31}H_{29}O_{14}$	625.1552	317.0700	14.02
Peonidin-3-(p-coumaroyl-glucoside)	Glucose	$C_{31}H_{29}O_{13}$	609.1603	301.0700	ND [v]
Malvidin-3-(p-coumaroyl-glucoside)	Glucose	$C_{32}H_{31}O_{14}$	639.1708	331.0800	ND
Delphinidin-3-(6″-acetyl-pyranoside)	Galactose/Glucose	$C_{23}H_{23}O_{13}$	507.1133	303.0500	ND
Cyanidin-3-(6″-acetyl-pyranoside)	Galactose/Glucose	$C_{23}H_{23}O_{12}$	491.1184	287.0550	ND
Petunidin-3-(6″-acetyl-pyranoside)	Galactose/Glucose	$C_{24}H_{25}O_{13}$	521.1289	317.0700	ND
Peonidin-3-(6″-acetyl-pyranoside)	Galactose/Glucose	$C_{24}H_{25}O_{12}$	505.1340	301.0700	ND
Malvidin-3-(6″-acetyl-pyranoside)	Galactose/Glucose	$C_{25}H_{27}O_{13}$	535.1446	331.0800	ND

[z] Compounds noted in different blueberry species, including rabbiteye, as reported in literature. [y] [M + H]+ = Molecular ion weight. Values corroborated by several literature sources, listed in Table 2. [x] MS/MS = Fragmented anthocyanidin molecular weight. Subsequently, this fragment ion (the backbone anthocyanidin), is free from the sugar moiety cleavage product. [w] R_t = retention time (minutes) from the Agilent 1200 HPLC. [v] ND = Compounds not detected in experimental samples. Theoretical molecular ion weights calculated based on literature.

Confirmation of anthocyanin identification was verified by the MS/MS scan of selected parent ion fragments and the sugar moiety molecular weight denoted in Table 1, using the MassHunter Workstation 6.00 software (Agilent, Santa Clara, CA, USA) [10]. Anthocyanin standards are difficult to find and the few that are available are expensive, so an extensive literature search was utilized to determine ions (Table 2).

Table 2. Anthocyanins in blueberries from the literature.

Anthocyanin	Literature [z]
Delphinidin-3-arabinoside	[3,10,19,28–32], [33], [34], [35], [36], [37], [38]
Cyanidin-3-arabinoside	[3,7,10,19,29–32], [33], [34], [35], [37], [38]
Petunidin-3-arabinoside	[7,10,19,28–32], [33], [35], [36], [37], [38]
Peonidin-3-arabinoside	[10], [26], [28,29,32], [33], [34], [35], [37], [38]
Malvidin-3-arabinoside	[7,10,19,28,29,31,32], [33], [34], [35], [36], [37]
Delphinidin-3-pyranoside	[3,7,10,19], [26], [28–32], [33], [34], [35], [36]
Cyanidin-3-pyranoside	[3,7,10,19], [26], [28–32], [33], [34], [35], [36,38]
Petunidin-3-pyranoside	[3,7,10,19], [26], [28–32], [33], [34], [35], [36,38]
Peonidin-3-pyranoside	[3,7,10,19], [26], [28–32], [33], [34], [35], [36,38]
Malvidin-3-pyranoside	[3,7,10,19], [26], [28–32,34], [35], [36], [38]

Table 2. *Cont.*

Anthocyanin	Literature [z]
Delphinidin-3-(p-coumaroyl-glucoside) [y]	[10]
Cyanidin-3-(p-coumaroyl-glucoside)	[10]
Petunidin-3-(p-coumaroyl-glucoside)	[10]
Peonidin-3-(p-coumaroyl-glucoside)	[10]
Malvidin-3-(p-coumaroyl-glucoside)	[10]
Delphinidin-3-(6″-acetyl-pyranoside)	[7,10,28,29,31], **[36]**
Cyanidin-3-(6″-acetyl-pyranoside)	[7,10,28,31]
Petunidin-3-(6″-acetyl-pyranoside)	[7,10,28,29,31]
Peonidin-3-(6″-acetyl-pyranoside)	[7,10,28,31]
Malvidin-3-(6″-acetyl-pyranoside)	[7,10,28,29,31], **[36]**

[z] References denoted by **bold font** indicate rabbiteye blueberry samples. [y] "p" indicates *para*.

2.3. Anthocyanidin Analysis

2.3.1. Extraction

Berry samples were analyzed for anthocyanidins using a modified UPLC method [10,28], as previously modified by Beaulieu et al. (2015) [39]. For the control samples, 10 g of raw berries were thawed and homogenized using a Tekmar Tissumizer (SDK-1810, IKA-Werke, Staufen, Germany) and a 2 g sample was lyophilized (VirTis Genesis 25ES, SP Scientific, Warminster, PA, USA). Press cake (PRC) was lyophilized in 40 g samples from each batch and a 2.5 g sample was utilized for extraction. Process step samples were 2 mL of lyophilized juice. All samples were allocated and weighed before being stored at −20 °C. Extraction of the juices and control was done using the 2 g frozen lyophilized powder in 2 mL of an extraction solvent comprised of 70:30:1 MeOH:H$_2$O:TFA. Press cake samples were extracted with 25 mL of extraction solvent and 2.5 g lyophilized powder. Each sample was vortexed for 15 s and left undisturbed for 60 min. Following the extraction, the samples were sonicated 20 min then centrifuged (IEC CL, International Equipment Company, Needham Heights, MA, USA) for 20 min at 7000 rpm and supernatant stored at −20 °C.

2.3.2. Acid Hydrolysis

The extracted samples (2 mL) were then hydrolyzed in a 4 mL vial, in which was mixed 200 μL of 12N HCl [28]. After purging vials with nitrogen gas, the vials were sealed and vortexed for 5 s and heated at 95 °C for 20 min. Immediately after heating, the samples were stored at −20 °C. Hydrolysates were thawed and filtered using a 0.2 μm polyvinylidene difluoride (PVDF) syringe filters, into autosampler vials.

2.3.3. UPLC Chromatography

Samples were analyzed on an Acquity ultra performance liquid chromatography system equipped with an ultra violet detector (UPLC–UV) using an Acquity BEH C18 column (50 mm × 2.1 mm × 1.7 μm) (Waters Corporation, Milford, MA, USA). The flow rate was 1.0 mL min^{-1}. The eluents consisted of acidified water with 3% phosphoric acid (A) and 100% acetonitrile (ACN) (B). The gradient started at 10% B, ramped to 20% B at 2 min, increased to 100% B at 2.1 min, then held at 100% B until 2.7 min, and returned to 10% B at 2.8 min. Detection was done with single-wavelength UV at 20 points min^{-1} at 525 nm. Anthocyanidin standards cyanidin, delphinidin, malvidin, peonidin, and petunidin (Chromadex, Santa Ana, CA, USA) were run at a concentration gradient of 0.001, 0.003, 0.010, 0.030 and 0.100 mg mL^{-1} ($r^2 \geq 0.995$) to report anthocyanidins (mg 100 g^{-1}).

2.4. Statistical Analysis

The experiment was repeated three times. Results are presented as mean ± standard deviation. Anthocyanin and anthocyanidin values obtained by HPLC–MS/MS and UPLC–UV experimentation are presented as the mean of three discreet batches, sampled three times, for a total of 9 data points for each treatment. Data were analyzed using analysis of variance (ANOVA), using the SAS (version 9.4; SAS Institute Inc., Cary, NC, USA). The means and standard deviations were compared using Tukey's studentized range method with $p \leq 0.05$.

3. Results and Discussion

3.1. Identification of Anthocyanins and Anthocyanidins

3.1.1. Anthocyanins

Anthocyanins were identified to determine the effects of processing on juice quality due to their heat sensitivity and as a marker of health benefits in blueberries [40]. Using literature to focus in and narrow down specific compounds, 10 major and 10 minor anthocyanins were evaluated, including 5 coumaroly-glucosides and 5 acetyl-pyranoside anthocyanins (Tables 1 and 2). Of these 20 compounds, 10 major and 3 minor anthocyanins were identified in the various blueberry juices and were used to visualize changes in juice quality through each processing step (Table 3).

Table 3. Average semi-quantitative LC–MS/MS anthocyanin peak area abundance in rabbiteye blueberry during various processing steps.

Treatment [z]	Del-3-ara [y]	Cya-3-ara	Pet-3-ara/Cya-3-pyr	Peo-3-ara
PRC	0.2 ± 0.05 d [x,w]	0.04 ± 0.03 d	0.2 ± 0.03 d	ND
Control	8.8 ± 1.8 a	8.4 ± 1.2 a	22.8 ± 2.9 a	5.1 ± 0.9 a
PJ	2.8 ± 0.5 b	2.4 ± 0.4 bc	7.0 ± 1.0 b	1.3 ± 0.2 bc
PPJ	2.7 ± 0.9 bc	2.6 ± 0.8 bc	7.3 ± 1.0 b	1.6 ± 0.5 ab
UF	3.6 ± 0.7 b	3.1 ± 0.6 b	8.3 ± 0.8 b	1.9 ± 0.2 b
UFP	1.7 ± 0.4 c	1.8 ± 0.4 c	4.4 ± 0.4 c	0.9 ± 0.2 c

	Mal-3-ara/Peo-3-pyr	Del-3-pyr	Pet-3-pyr	Mal-3-pyr
PRC	0.3 ± 0.1 d	0.5 ± 0.09 d	0.3 ± 0.03 d	0.5 ± 0.1 d
Control	49.2 ± 8.3 a	20.5 ± 4.1 a	18.8 ± 3.8 a	49.5 ± 9.7 a
PJ	16.5 ± 2.8 bc	7.1 ± 1.5 b	7.1 ± 1.6 b	24.7 ± 4.4 bc
PPJ	17.7 ± 5.6 b	6.0 ± 1.7 bc	7.0 ± 2.2 b	25.6 ± 7.9 b
UF	22.1 ± 2.3 b	8.4 ± 1.3 b	8.7 ± 1.9 b	30.5 ± 4.6 b
UFP	12.2 ± 3.6 c	4.0 ± 0.6 c	4.3 ± 0.9 c	18.2 ± 4.6 c

	Del-3-cou	Cya-3-cou	Pet-3-cou	Total Abundance
PRC	0.05 ± 0.01 c	ND	0.1 ± 0.05 c	2.2 ± 0.4 d
Control	0.5 ± 0.1 b	0.3 ± 0.05 b	1.4 ± 0.5 b	185.2 ± 34.0 a
PJ	1.1 ± 0.3 a	0.4 ± 0.1 a	2.1 ± 0.7 ab	72.5 ± 13.4 b
PPJ	0.9 ± 0.4 a	0.4 ± 0.1 ab	2.2 ± 1.1 a	74.0 ± 21.2 b
UF	1.1 ± 0.4 a	0.5 ± 0.2 a	2.9 ± 1.1 a	85.2 ± 21.2 b
UFP	1.0 ± 0.3 a	0.4 ± 0.08 ab	2.1 ± 0.4 ab	50.9 ± 11.5 c

[z] PRC = press cake; PJ = pressed not filtered juice; PPJ = pasteurized pressed unfiltered juice; UF = ultrafiltered juice; UFP = ultrafiltered pasteurized juice. [y] Del-3-ara = delphinidin-3-arabinoside; Cya-3-ara = cyanidin-3-arabinoside; Pet-3-ara/Cya-3-pyr = petunidin-3-arabinoside/cyanidin-3-pyranoside; Peo-3-ara = peonidin-3-arabinoside; Mal-3-ara/Peo-3-pyr = malvidin-3-arabinoside/peonidin-3-pyranoside; Del-3-pyr = delphinidin-3-pyranoside; Pet-3-pyr = petunidin-3-pyranoside; Mal-3-pyr = malvidin-3-pyranoside; Del-3-cou = delphinidin-3-(p-coumaroyl-glucoside); Cya-3-cou = cyanidin-3-(p-coumaroyl-glucoside); Pet-3-cou = petunidin-3-(p-coumaroyl-glucoside). [x] Abundance values are reported as average ion counts min^{-1} ± standard deviations, ×100,000, according to information presented per compound in Table 1. [w] Significant differences (Tukey's method with $p < 0.05\%$) per parameter are designated by letters in each column. Means not connected by the same letter are significantly different at $p < 0.05$.

Conformation of anthocyanins was determined by the anthocyanin molecular weight ion $[M + H]^+$ and the backbone anthocyanidin molecular weight ion (MS/MS), as verified in literature [10,41]. With the exception of peonidin-3-arabinoside and cyanidin-3-(p-coumaroyl-glucoside) in the press cake, all 13 major and minor anthocyanidins were identified in each processing step. This corroborates with other studies identifying anthocyanins in blueberries, including RAB berries (Table 2) [34–36,38]. Several studies on southern highbush (SHB) and northern highbush (NHB) blueberries, as well as bilberries, identified minor 5 acetyl-pyranoside anthocyanins, which were not identified in this study (Table 2). This may be due to varietal and species differences between RAB berries and other blueberry types [36,42]. A "Tifblue" RAB juice study by Srivastava et al. (2007) [26] created juice by thawing, boiling and blending the berries in a household blender before treating them with a pectinase enzyme. The berry slurry was then centrifuged and batch pasteurized to 85 °C for 2 min before bottling. Results of their bench top experiment identified 8 anthocyanins using HPLC–UV (Table 2). Previously, glucoside and galactoside anthocyanins were isolated in RAB berries [26], but they did not identify as many arabinoside compounds as our experiment. An LC–MS overlay of the major anthocyanin abundances in a raw berry sample (control) chromatogram shows how closely the compounds elute relative to each other and how MS/MS can help to differentiate some of the different compounds (Figure 1).

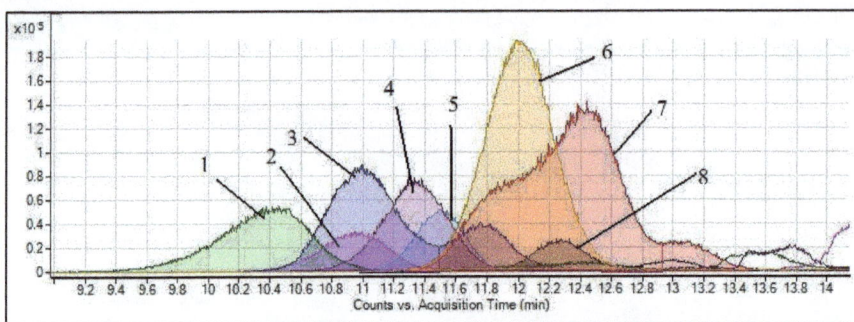

Figure 1.   Overlay of [M + H]$^+$ ions of major anthocyanins in raw berry sample. (1) Delphinidin-3-pyrannoside (m/z 465 [M + H]$^+$); (2) Delphinidin-3-arabinoside (m/z 435 [M + H]$^+$); (3) Cyanidin-3-pyrannoside (m/z 449 [M + H]$^+$); (4) Petunidin-3-pyrannoside (m/z 479 [M + H]$^+$); (5) Cyanidin-3-arabinoside (m/z 419 [M + H]$^+$); (6) Malvidin-3-pyrannoside (m/z 493 [M + H]$^+$); (7) Malvidin-3-arabinoside/Peonidin-3-pyrannoside (m/z 463 [M + H]$^+$); (8) Peonidin-3-arabinoside (m/z 433 [M + H]$^+$).

In the major anthocyanins identified, separation issues were identified between the glucoside- and galactoside-containing anthocyanins. These two sugar moieties proved difficult to separate the 6-carbon structures with identical [M + H]$^+$ and MS/MS ions as well as similar elution times (Figure 1, Peaks 1, 3, 4 and 6). Peonidin-3-arabinoside was not separated from cyanidin-3-pyranoside as well as malvidin-3-arabinoside from peonidin-3-pyranoside (Figure 1, Peaks 7 and 8). These compounds' elution times, almost identical [M + H]$^+$ fragment ions MW, and difficulty to separate pyranoside compounds made separation impractical. In the past, extensive LC method development for peak separation has been reported [31]. Therefore, the unresolved co-eluding peaks were evaluated as one peak response and not individual anthocyanins (Table 3).

As part of the complexing and/or co-pigmentation of anthocyanins, they can also form esters with hydroxycinnamates and organic acids [41]. The lesser anthocyanins may consist of acetoly, malonoyl and coumaroyl compounds [10]. In this study, 5 coumaroly-glucosides and 5 acetyl-pyranoside anthocyanins were searched for and three minor coumaroyl anthocyanin compounds were identified using MS/MS ions, as confirmed in literature (Table 1) [30,41,43]. The compounds were delphinidin-3-(p-coumaroyl-glucoside), cyanidin-3-(p-coumaroyl-glucoside), and petunidin-3-(p-coumaroyl-glucoside) which, to the best of our knowledge, have not been reported in RAB blueberries (Figure 2). These three coumaroyl anthocyanins are reported in grapes, radishes, red cabbage and black carrots [41,44]. A possibility for not identifying the other acetyl-pyranoside anthocyanins as identified by other studies in RAB berry samples may be due to the inadequate LC separation before MS detection of the compounds: including phase gradients, length of gradient, and run times. Figure 3 illustrates the MS/MS scan of the 3-coumaroyl compounds and the ion fragments of each compound. Further LC–MS/MS method development is apparently needed to better separate and further confirm identification of the minor anthocyanins in RAB berries, which could provide further tools to monitor processing variables and juice quality.

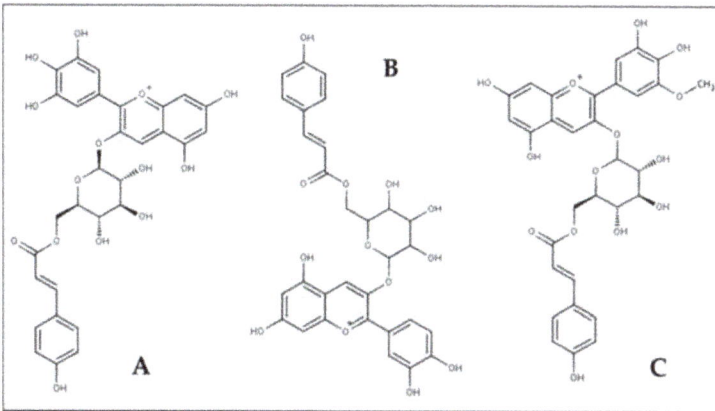

Figure 2. Chemical structure of delphinidin-3-(p-coumaroyl-glucoside) (**A**); cyanidin-3-(p-coumaroyl-glucoside) (**B**) and petunidin-3-(p-coumaroyl-glucoside (**C**). Figure adapted from (Neveu et al., 2012) [43].

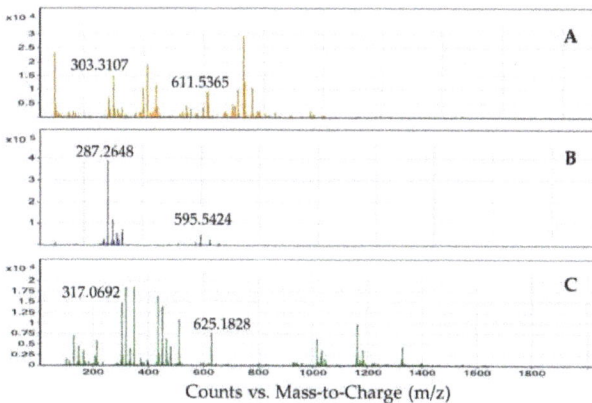

Figure 3. MS/MS spectrum scan of 3 identified minor anthocyanins in unfiltered rabbiteye blueberry juice: delphinidin-3-(p-coumaroyl-glucoside), top plate (**A**); cyanidin-3-(p-coumaroyl-glucoside), middle plate (**B**) and petunidin-3-(p-coumaroyl-glucoside) bottom plate (**C**).

3.1.2. Anthocyanidins

The five major anthocyanidins found in blueberries (cyanidin, delphinidin, malvidin, peonidin and petunidin) were all identified in every step of the juice process. A chromatogram of the anthocyanidins found in a press cake sample illustrates the elution order for the anthocyanidins in RAB blueberries (Figure 4). Acid hydrolysis of anthocyanins produces anthocyanidins which are identified with UV detectors. Sugar moieties have extremely similar molecular weights which can complicate identification [28]. By removing the glycosides from the molecule, the anthocyanidin backbones can be evaluated and precisely quantified to illustrate the classes of anthocyanin molecules present in the sample.

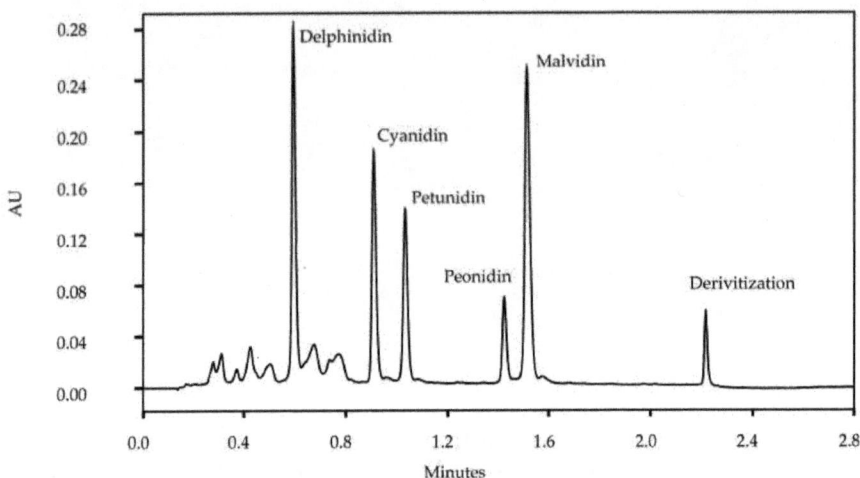

Figure 4. Anthocyanidins found in rabbiteye blueberry press cake (PRC) as detected by HPLC–UV.

3.2. Changes of Anthocyanins during Blueberry Juice Processing

Anthocyanin amounts were measured to determine the effects of processing on juice quality. Using the semi-quantitative LC–MS/MS peak area abundance data to compare anthocyanins, the most abundant major anthocyanin identified in the control rabbiteye sample was malvidin-3-pyranoside (Table 3). Malvidin-3-galactoside was the most abundant anthocyanin found in NHB blueberries [37]. Although peak resolution did not allow for the separation of the two 6 carbon sugars, herein, malvidin-3-galactoside was included in the recovery of malvidin-3-pyranoside. The least abundant major anthocyanin identified was peonidin-3-arabinoside. Comparing these results to a study by Srivastava et al. (2007) [26], malvidin-3-pyranoside was also found to be the largest anthocyanin recovered; however, they concluded that cyanidin-3-pyranoside was the least abundant anthocyanin in "Tifblue" RAB juice.

After pressing, a large percentage of anthocyanins were not transferred to the PJ from the whole berry (control). With one exception, there was an approximate 70% decrease between control and PJ in all anthocyanins (Table 3). However, malvidin-3-pyranoside (Mal-3-pyr) only had a 50% loss. Peonidin-3-arabinoside (Peo-3-ara) had the greatest loss (74%), decreasing from $5.1 \pm 0.9 \times 10^5$ to $1.3 \pm 0.2 \times 10^5$ counts. The anthocyanin decreases between the PJ and the pressed pasteurized unfiltered juice (PPJ) were minimal and not significantly different. This may be explained by the increased heat stability caused by possible polymerization and/or co-pigmentation of the anthocyanins. Co-pigmentation of anthocyanins with other anthocyanins and polyphenols increases the molecule's heat stability [45].

After the initial decrease from pressing the berries, the second greatest decrease in anthocyanins occurred when ultrafiltered juice was pasteurized (UFP). The major anthocyanins significantly decreased 40% to 53% after pasteurization in the UFP samples (Table 3). Major anthocyanins containing delphinidin had the greatest degradation, with delphinidin-3-pyranoside (Del-3-pyr) decreasing from $8.4 \pm 1.3 \times 10^5$ to $4.0 \pm 0.6 \times 10^5$ counts. Delphinidin was less stable in elevated temperatures which resulted in greater decreases in relation to the other anthocyanins due to its greater heat lability [26].

The minor anthocyanin trends after pressing were opposite compared to the major anthocyanins. Both cyanidin-3-(p-coumaroyl-glucoside) (Cya-3-cou) and petunidin-3-(p-coumaroyl-glucoside) (Pet-3-cou) increased (occasionally significantly) ~33% in PJ samples from their initial control levels. Delphinidin-3-(p-coumaroyl-glucoside) (Del-3-cou) significantly increased 50% from $0.5 \pm 0.1 \times 10^5$ to $1.1 \pm 0.3 \times 10^5$ counts after pressing (Table 3). These increases may be due to the heat degradation process by which anthocyanins break down into chalcone structures which can then undergo transformation into a coumarin glucoside [46]. One possible cause of the recoveries for the minor anthocyanins through processing involves the acylation of the anthocyanin molecule and its ability to increase stability by preventing hydration when exposed to pH changes and heat [44,47]. On the other hand, the greater anthocyanin degradation in the UFP juices as compared to the PPJ juices may be directly related to the role that co-pigmentation and anthocyanin polymers and/or polymerization play in the heat stability of the compounds themselves. Ultrafiltration removes many of the higher molecular weight co-pigments and polymers from the juice. It can be assumed that the larger the polymer or co-pigment, the greater the increase in heat stability of the compounds included in the polymer [45].

There was a significant 61% decrease in total anthocyanins between the control and PJ samples in the juice processing steps. The control berries contained $185.2 \pm 34.0 \times 10^5$ counts min^{-1}, while the PJ juice contained only $72.5 \pm 13.4 \times 10^5$ counts (Table 3). This was expected due to the large percentage of anthocyanins that remain in the press cake [48]. There was an insignificant concentration increase in the amounts of anthocyanin content of the UF samples, as compared to the PJ (Table 3). A concentrating effect regarding monomer anthocyanins has been observed in grape juice filtration using varying sizes of membranes [49]. Utilization of filtration to increase the concentration of beneficial compounds in juices is a concept that is being explored in other fruits and can increase juice nutritional value [50–53]. Polyphenolics in the retentate (recirculated to the 100 L hopper) from the filtration process were not evaluated in this experiment. However, since sugars from controls (12.7 °Brix) increased in the residual retentate (16 °Brix), the residual retentate presumably accumulates other possible value-added by-products that the producer may capitalize upon, in addition to the press cake. Removal of the anthocyanin polymers could be utilized as more stable natural food colorings [51]. Ultrafiltered juices (UFP) had significantly greater anthocyanin degradation due to pasteurization compared with the PPJ juice. A 40% decrease in total anthocyanins after pasteurization of UF juices, similar to microfiltered pomegranate juice [54], may be attributed to anthocyanin degradation from pasteurization due to the removal of the co-pigmentation of the anthocyanins by ultrafiltration [55].

Significant anthocyanin losses have also occurred in NHB blueberry juice and products [19,37,48,56,57]. There was 58.8% loss of monomeric anthocyanins in enzyme-treated, non-clarified, batch-pasteurized NHB blueberry juice prepared from frozen blanched berries, but only 27.5% loss in centrifuge-clarified juice [48]. A slightly different process using thawed, blanched NHB blueberries treated with pectinase, pressed, then clarified and followed by HTST (90 °C) pasteurization resulted in higher losses of 77.6% and 84.3% anthocyanin glycosides and anthocyanins, respectively [19]. There was a 68.0% total anthocyanin loss in juice created with individually quick frozen (IQF) NHB blueberries that were partially thawed, heated to 43 °C, treated with pectinase, pressed, then HTST pasteurized (90 °C) [37]. In puree/juice made with NHB berries that were manually crushed and machine-juiced, not filtered or clarified, then batch-pasteurized at 92 °C, there was a substantial 95.7% loss of total anthocyanins [57]. In addition, there was a 76.2% and 79.1% anthocyanin loss in Powderblue and Tifblue RAB blueberry varieties, respectively, that were thawed

12 h at 5 °C, hot water blanched, homogenized in a blender, pectinase added, centrifuged, then batch-pasteurized (85 °C) [26]. Overall, significant anthocyanin "losses" occur in processed blueberry juices and, herein, we did not prevent losses using a heated mash and UF with NFC juices.

The MS/MS PRC samples had unusually low anthocyanin peak area abundance amounts reported overall, compared to the remaining processing steps (Table 3). However, the low reports of anthocyanins in the PRC were likely due to problems with the extraction method itself (discussed below in Section 3.3.1) and did not reflect the actual amounts of anthocyanins found in the PRC. As discussed below, anthocyanidin recoveries from the PRC support the supposition that our extraction protocols for MS/MS PRC samples versus the liquid UPLC–UV samples were not comparatively effective, because the majority of anthocyanidins indeed remained in the PRC.

3.3. Changes of Anthocyanidins during Blueberry Juice Processing

3.3.1. Individual Anthocyanidins

To better define the changes caused by juice processing on anthocyanins, anthocyanidins were also evaluated. Malvidin (26.6 mg 100 g^{-1}) and petunidin (25.9 mg 100 g^{-1}) were the most abundant anthocyanidins in the control samples, while peonidin (4.1 mg 100 g^{-1}) was the least abundant. Markedly higher amounts of anthocyanidins detected in the press cake are due to the majority of anthocyanins being located in the skins of the berries [37]. Press cake samples contained 3-fold greater total anthocyanidins (265.6 mg 100 g^{-1}) than the control (85.1 mg 100 g^{-1}). Since berry skins account for 10% to 20% of the berry weight (13.0 ± 0.6% herein [58]) and as the main location for anthocyanins in the berry, the majority of anthocyanins were left behind in the press cake, as previously demonstrated [19,37,48]. Cyanidin was the most abundant anthocyanidin (77.7 mg 100 g^{-1}) recovered from PRC.

To further illustrate changes caused by juice processing, percent change was calculated for individual anthocyanidins at each step as compared against the frozen berry controls. Delphinidin retained the most from the control with 48.7% continuing into the PJ juice; however, only 22.5% of petunidin was transferred (Figure 5). After pasteurization of PJ juice, cyanidin was retained 41.8% from the initial control, while petunidin was retained only 20.3%. In UF juices, delphinidin again was retained the most in the initial control anthocyanidins (34.2%) while petunidin was retained the least (16.7%) (Figure 5). The same pattern was observed in UFP juices with delphinidin (34.9%) and petunidin (14.1%). The percent change for total anthocyanidins summarizes the data further. At pressing, 32.9% of the control total anthocyanidins were transferred to the resulting PJ juice, resulting in a 67.1% loss of total anthocyanidins (Figure 5). After pasteurization, the unfiltered pasteurized juice (PPJ) contained 29.7% of the starting control total anthocyanidins. Further processing of the pressed juice (PJ) created the UF juice which had 24.5% of the initial control berry anthocyanidins, hence increasing the total anthocyanidin loss to 75.5%. The UFP, lastly, had 21.7% of the total anthocyanins that the control berries started with, rendering losses of 78.3% (Figure 5).

3.3.2. Total Anthocyanidins

Juice producers are more concerned with total anthocyanin changes then individual compounds and the relation to how steps in the juice process affect the compounds that change clarity and cause sedimentation. Since the anthocyanins are only semi-quantified and the PRC samples were not adequately extracted for a good representation of the individual MS/MS amounts, the total anthocyanins could not be utilized to track the percent of anthocyanins that proceeded forward from the whole berry. This was, however, calculated with the UPLC quantified anthocyanidin data. To determine how much of the juice and press cake came from the raw berries for each sample, relative percentages were determined by utilizing previously collected data from pilot presses [58]. The percentage of the whole berries broken down into relative free juice and press cake amounts were 74.0 ± 1.0% and 13.0 ± 0.6%, respectively (note that through experimental batch-like transfers, UF hopper residuals and measuring, $13 \pm 0.5\%$ juice was also lost, which was added to the value to calculate "total juice" versus cake [58]). The percent of anthocyanidins that remained in the juice and press cake after pressing, as well as percentage lost after each processing step were calculated using these estimations (Figure 6). The sequential degradation or loss/polymerization of anthocyanins begins with loss to pressing, which removed the majority of anthocyanidins. There was a 40.6% loss of anthocyanidins to the press cake and 35.1% was lost to mashing and juice loss in the press cloth and equipment. Of the total anthocyanidins in raw control berries, only 24.3% were transferred to the juice (PJ) (Figure 6). Thereafter, juice that was not filtered lost another 9.6% through pasteurization (PPJ). This translates to 29.7% of the original control amount of anthocyanidins being transferred to a non-filtered pasteurized juice product (not counting the effectually concentrated PRC loss). By filtering juice, an additional 25.4% of the anthocyanidins were removed from PJ samples before pasteurization (UF) (Figure 6). Pasteurization of ultrafiltered juice (UFP) removed an additional 12.0% of anthocyanidins in the final juice product. The UFP contained roughly 21.6% of the raw berry control anthocyanidins after processing.

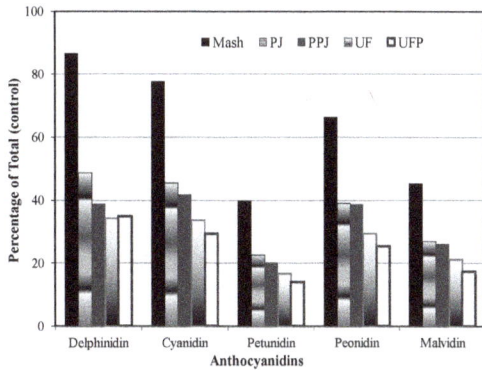

Figure 5. Percentage of control of anthocyanidins in rabbiteye blueberry juice processing steps. Mash = crude mash following the steam-jacketed kettle (prior to enzyme treatment and hydraulic pressing); PJ = pressed not filtered juice; PPJ = pasteurized pressed unfiltered juice; UF = ultrafiltration-filtered juice; UFP = filtered pasteurized juice.

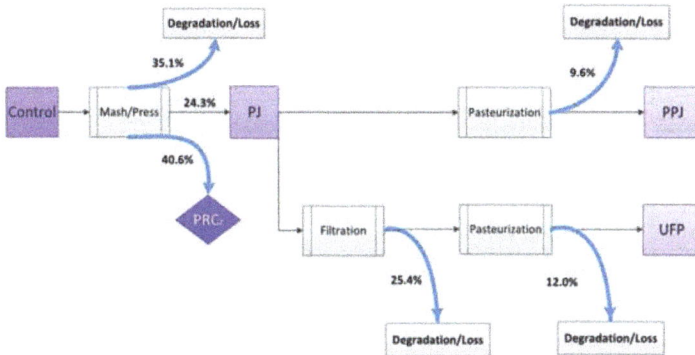

Figure 6. Percent anthocyanidin loss through unfiltered and ultrafiltered rabbiteye blueberry juice processing. PRC = press cake; PJ = pressed not filtered juice; PPJ = pasteurized pressed unfiltered juice; UFP = filtered pasteurized juice. Filtration box specifically refers to ultrafiltration (UF), as described in the M&M.

Literature on anthocyanidins in blueberries give varying results in comparison to our study. A contributing factor is the difference in calculating total anthocyanidins by summation of individual anthocyanidin amounts compared to using the pH differential method which calculates total anthocyanidins as one group of compounds based on UV spectra [59,60]. In general, blueberry results reported from a spectrophotometric method are higher in the same sample than results from methods quantifying the individual anthocyanidins [19,30,61]. Yet, the reverse situation has also been reported [62]. Although the pH differential method versus LC analyses can deliver differing results from the same sample, blueberry data tend to be closely correlated, which indicates that both approaches are reliable [60,61,63]. The loss to press cake is similar to the percentages found in the literature [19,37]. Highlighting the amount of anthocyanidins remaining in the press cake again confirms that economic opportunities abound for producers to further utilize their waste stream. Minimizing waste from the juice production scheme should be accomplished as value-added polyphenolic products could be developed from the press cake. Greener technologies, like environmentally friendly hot water extraction, is an option for removing the remaining beneficial compounds from press cake [64]. The seeds and skins which remain in the press cake can be separated and utilized for seed oil, natural coloring, and confectionary products [15], creating sustainable and unique perspectives for marketing niche products.

4. Conclusions

Ten major and three minor anthocyanins were identified in RAB blueberry juice processing steps. There were three minor anthocyanins isolated, which have not been reported previously in RAB blueberries. All five of the major anthocyanidin classes found in blueberry were identified as well. Comparing the trends of the individual anthocyanins, the major individual compounds all significantly decreased after pressing and pasteurization; the three minor anthocyanins, while increased slightly, with significant changes varying based on the process steps through juice processing. Acylation of the minor anthocyanins may increase stability when exposed to pH changes and heat, and this may explain why different trends were observed in the minor coumaroyl anthocyanins compared with the major anthocyanins. Raw control berries and press cake contained the highest anthocyanidin contents. Pressed juice only contained 24.3% of the anthocyanidins transferred from raw control berry quantities. Unfiltered juices only had 9.6% loss of anthocyanidins resulting from pasteurization, while ultrafiltered juices had 12.0% loss from pasteurization, in addition to the 25.4% anthocyanidin loss from the ultrafiltration step. Herein, we too realized that ultrafiltration concentrated monomer anthocyanins slightly and ultrafiltration should be further evaluated and optimized to increase anthocyanins in NFC juices. These findings also contribute to the value and interest of press cake for use in other food and non-food products as value-added ingredients and product quality to boost health benefits. In juice processing, the more steps in the process, the greater the loss of anthocyanins in the end product. Nonetheless, further studies evaluating the storage effects on juice quality between unfiltered and filtered NFC juices, taking into account lower molecular weight anthocyanins and polymers associated with increased health benefits, would contribute to the important role of filtration in juice processing.

Acknowledgments: We would like to thank Donna Marshall-Shaw and Lavonne Stringer at the Thad Cochran Southern Horticultural Laboratory, USDA, ARS, in Poplarville, for their help with the blueberries.

Author Contributions: The corresponding author, John C. Beaulieu, had equal and shared contribution as the senior author in this work; John C. Beaulieu conceived and designed the experiments; Rebecca E. Stein-Chisholm and John C. Beaulieu performed the experiments; Steven W. Lloyd refined the UPLC method and mentored Rebecca E. Stein-Chisholm to run samples and interpret data; Rebecca E. Stein-Chisholm initiated and facilitated the LC MS/MS analysis which Casey C. Grimm helped develop, and mentored Rebecca E. Stein-Chisholm develop an in-house anthocyanin library as she ran samples and interpreted the data; Rebecca E. Stein-Chisholm performed much of the lab work, along with John C. Beaulieu and Steven W. Lloyd; Rebecca E. Stein-Chisholm analyzed the data; John C. Beaulieu and Casey C. Grimm contributed reagents/materials/equipment; Rebecca E. Stein-Chisholm and John C. Beaulieu co-wrote the paper..

Conflicts of Interest: The authors declare no conflict of interest.

References

1. Kalt, W.; Dufour, D. Health functionality of blueberries. *HortTechnology* **1997**, *7*, 216–221.
2. Soto-Vaca, A.; Gutierrez, A.; Losso, J.N.; Xu, Z.; Finley, J.W. Evolution of phenolic compounds from color and flavor problems to health benefits. *J. Agric. Food Chem.* **2012**, *60*, 6658–6677. [CrossRef] [PubMed]
3. Zhang, Z.; Kou, X.; Fugal, K.; McLaughlin, J. Comparison of HPLC methods for determination of anthocyanins and anthocyanidins in bilberry extracts. *J. Agric. Food Chem.* **2004**, *52*, 688–691. [CrossRef] [PubMed]
4. Gould, K.; Davies, K.M.; Winefield, C. *Anthocyanins: Biosynthesis, Functions, and Applications*, 1st ed.; Springer: New York, NY, USA, 2009; p. 325.
5. Pina, F.; Melo, M.J.; Laia, C.A.T.; Parola, A.J.; Lima, J.C. Chemistry and applications of flavylium compounds: A handful of colours. *Chem. Soc. Rev.* **2012**, *41*, 869–908. [CrossRef] [PubMed]
6. Wu, X.; Beecher, G.R.; Holden, J.M.; Haytowitz, D.B.; Gebhardt, S.E.; Prior, R.L. Concentrations of anthocyanins in common foods in the United States and estimation of normal consumption. *J. Agric. Food Chem.* **2006**, *54*, 4069–4075. [CrossRef] [PubMed]
7. Gao, L.; Mazza, G. Quantitation and distribution of simple and acylated anthocyanins and other phenolics in blueberries. *J. Food Sci.* **1994**, *59*, 1057–1059. [CrossRef]
8. Baxter, H.; Harborne, J.B.; Moss, G.P. *Phytochemical Dictionary: A Handbook of Bioactive Compounds from Plants*, 2nd ed.; CRC Press: Boca Raton, FL, USA, 1998; p. 976.
9. Routray, W.; Orsat, V. Blueberries and their anthocyanins: Factors affecting biosynthesis and properties. *Compr. Rev. Food Sci. Food Saf.* **2011**, *10*, 303–320. [CrossRef]
10. Barnes, J.S.; Nguyen, H.P.; Shen, S.; Schug, K.A. General method for extraction of blueberry anthocyanins and identification using high performance liquid chromatography–electrospray ionization-ion trap-time of flight-mass spectrometry. *J. Chromatogr. A* **2009**, *1216*, 4728–4735. [CrossRef] [PubMed]
11. Tropp, D. Why Local Food Matters: The Rising Importance of Locally-grown Food in the U.S. Food System. In Proceedings of the National Association of Counties Legislative Conference, Washington, DC, USA, 28 July 2014.
12. Barkla, C. Fruit juices. In *Market News Service*; International Trade Centre (A joint agency of the World Trade Organization and the United Nations): Geneva, Switzerland, 2011; pp. 1–21.
13. AIJN. *Fruit Juice Matters 2017 Report*; European Fruit Juice Association: Brussels, Belgum, 2017.
14. Rohan, M. The markets for soft drinks and fruit juices change as a constant—Focusing on health. In *Food and Drink Business Europe*; Rohan, M., Ed.; Premier Publishing Ltd.: Dublin, Ireland, 2013.
15. Bates, R.P.; Morris, J.R.; Crandall, P.G. Principles and Practices of Small and Medium Scale Fruit Juice Processing. FAO Agricultural Services Bulletin 2001. Available online: http://www.fao.org/docrep/005/y2515e/y2515e00.htm (accessed on 11 November 2017).
16. Nikdel, S.; Chen, C.S.; Parish, M.E.; MacKellar, D.G.; Friedrich, L.M. Pasteurization of citrus juice with microwave energy in a continuous-flow unit. *J. Agric. Food Chem.* **1993**, *41*, 2116–2119. [CrossRef]
17. Perera, C.O.; Smith, B. Technology of processing of horticultural crops. In *Handbook of Farm, Dairy, and Food Machinery Engineering*, 2nd ed.; Kutz, M., Ed.; Academic Press: Cambridge, MA, USA, 2013; pp. 259–315.
18. Brambilla, A.; Maffi, D.; Rizzolo, A. Study of the influence of berry-blanching on syneresis in blueberry purées. *Procedia Food Sci.* **2011**, *1*, 1502–1508. [CrossRef]
19. Lee, J.; Durst, R.W.; Wrolstad, R.E. Impact of juice processing on blueberry anthocyanins and polyphenolics: Comparison of two pretreatments. *J. Food Sci.* **2002**, *67*, 1660–1667. [CrossRef]
20. Brambilla, A.; Lo Scalzo, R.; Bertolo, G.; Torreggiani, D. Steam-blanched highbush blueberry (*Vaccinium corymbosum* L.) juice: Phenolic profile and antioxidant capacity in relation to cultivar selection. *J. Agric. Food Chem.* **2008**, *56*, 2643–2648. [CrossRef] [PubMed]

21. Landbo, A.-K.; Kaack, K.; Meyer, A.S. Statistically designed two step response surface optimization of enzymatic prepress treatment to increase juice yield and lower turbidity of elderberry juice. *Innov. Food Sci. Emerg. Technol.* **2007**, *8*, 135–142. [CrossRef]
22. Alper, N.; Bahceci, K.S.; Acar, J. Influence of processing and pasteurization on color values and total phenolic compounds of pomegranate juice. *J. Food Process. Preserv.* **2005**, *29*, 357–368. [CrossRef]
23. Brooks, R.M.; Olmo, H.P. *The Brooks and Olmo Register of Fruit & Nut Varieties*; ASHS Press: Alexandria, VA, USA, 1997.
24. Marshall, D.A.; Spiers, J.M.; Braswell, J.H. Splitting severity among rabbiteye (*Vaccinium ashei* 'Reade') blueberry cultivars in Mississippi and Louisiana. *Intl. J. Fruit Sci.* **2006**, *6*, 77–81. [CrossRef]
25. USDA. *Germplasm Resources Information Network (GRIN) Online Database*; United States Department of Agriculture, Agricultural Research Service, National Genetic Resources Program, Ag Data Commons, 2014. Available online: http://dx.doi.org/10.15482/USDA.ADC/1212393 (accessed on 11 November 2017).
26. Srivastava, A.; Akoh, C.C.; Yi, W.; Fischer, J.; Krewer, G. Effect of storage conditions on the biological activity of phenolic compounds of blueberry extract packed in glass bottles. *J. Agric. Food Chem.* **2007**, *55*, 2705–2713. [CrossRef] [PubMed]
27. Beaulieu, J.C.; Stein-Chisholm, R.E.; Lloyd, S.W.; Bett-Garber, K.L.; Grimm, C.C.; Watson, M.A.; Lea, J.M. Volatile, anthocyanidin, quality and sensory changes in rabbiteye blueberry from whole fruit through pilot plant juice processing. *J. Sci. Food Agric.* **2017**, *97*, 469–478. [CrossRef] [PubMed]
28. Hynes, M.; Aubin, A. *Acquity UPLC for the Rapid Analysis of Anthocyainidins in Berries*; Waters Corporation: Milford, MA, USA, 2006.
29. Cho, M.J.; Howard, L.R.; Prior, R.L.; Clark, J.R. Flavonoid glycosides and antioxidant capacity of various blackberry, blueberry and red grape genotypes determined by high-performance liquid chromatography/mass spectrometry. *J. Sci. Food Agric.* **2004**, *84*, 1771–1782. [CrossRef]
30. Fanali, C.; Dugo, L.; D'Orazio, G.; Lirangi, M.; Dachà, M.; Dugo, P.; Mondello, L. Analysis of anthocyanins in commercial fruit juices by using nano-liquid chromatography-electrospray-mass spectrometry and high-performance liquid chromatography with UV-vis detector. *J. Sep. Sci.* **2011**, *34*, 150–159. [CrossRef] [PubMed]
31. Kalt, W.; McDonald, J.E.; Ricker, R.D.; Lu, X. Anthocyanin content and profile within and among blueberry species. *Can. J. Plant Sci.* **1999**, *79*, 617–623. [CrossRef]
32. Lee, J.; Finn, C.E.; Wrolstad, R.E. Comparison of anthocyanin pigment and other phenolic compounds of *Vaccinium membranaceum* and *Vaccinium ovatum* native to the pacific northwest of North America. *J. Agric. Food Chem.* **2004**, *52*, 7039–7044. [CrossRef] [PubMed]
33. Nakajima, J.-I.; Tanaka, I.; Seo, S.; Yamazaki, M.; Saito, K. LC/PDA/ESI-MS profiling and radical scavenging activity of anthocyanins in various berries. *J. Biomed. Biotechnol.* **2004**, *5*, 241–247. [CrossRef] [PubMed]
34. Lohachoompol, V.; Mulholland, M.; Srzednicki, G.; Craske, J. Determination of anthocyanins in various cultivars of highbush and rabbiteye blueberries. *Food Chem.* **2008**, *111*, 249–254. [CrossRef]
35. Prior, R.L.; Lazarus, S.A.; Cao, G.; Muccitelli, H.; Hammerstone, J.F. Identification of procyanidins and anthocyanins in blueberries and cranberries (*Vaccinium* spp.) using high-performance liquid chromatography/mass spectrometry. *J. Agric. Food Chem.* **2001**, *49*, 1270–1276. [CrossRef] [PubMed]
36. Skrede, G.; Wrolstad, R.E.; Durst, R.W. Changes in anthocyanins and polyphenolics during juice processing of highbush blueberries (*Vaccinium corymbosum* L.). *J. Food Sci.* **2000**, *65*, 357–364. [CrossRef]
37. Yoshimura, Y.; Enomoto, H.; Moriyama, T.; Kawamura, Y.; Setou, M.; Zaima, N. Visualization of anthocyanin species in rabbiteye blueberry *Vaccinium ashei* by matrix-assisted laser desorption/ionization imaging mass spectrometry. *Anal. Bioanal. Chem.* **2012**, *403*, 1885–1895. [CrossRef] [PubMed]
38. Wu, X.; Prior, R.L. Systematic identification and characterization of anthocyanins by HPLC-ESI-MS/MS in common foods in the United States: Fruits and berries. *J. Agric. Food Chem.* **2005**, *53*, 2589–2599. [CrossRef] [PubMed]
39. Beaulieu, J.C.; Lloyd, S.W.; Preece, J.E.; Moersfelder, J.W.; Stein-Chisholm, R.E.; Obando-Ulloa, J.M. Physicochemical properties and aroma volatile profiles in a diverse collection of California-grown pomegranate (*Punica granatum* L.) germplasm. *Food Chem.* **2015**, *181*, 354–364. [CrossRef] [PubMed]
40. Havlíková, L.; Míková, K. Heat stability of anthocyanins. *Z. Lebensm. Forch.* **1985**, *181*, 427–432. [CrossRef]

41. Burns, J.; Mullen, W.; Landrault, N.; Teissedre, P.-L.; Lean, M.E.J.; Crozier, A. Variations in the profile and content of anthocyanins in wines made from Cabernet Sauvignon and hybrid grapes. *J. Agric. Food Chem.* **2002**, *50*, 4096–4102. [CrossRef] [PubMed]

42. Howard, L.R.; Clark, J.R.; Brownmiller, C. Antioxidant capacity and phenolic content in blueberries as affected by genotype and growing season. *J. Sci. Food Agric.* **2003**, *83*, 1238–1247. [CrossRef]

43. Neveu, V.; Perez-Jimenez, J.; Vos, F.; Crespy, V.; du Chaffaut, L.; Mennen, L.; Knox, C.; Elisner, R.; Cruz, J.; Wishart, D.; et al. Phenol-Explorer: An online comprehensive database on polyphenol contents in foods. *Database* **2010**, *2010*, 1–9. [CrossRef] [PubMed]

44. Giusti, M.M.; Wrolstad, R.E. Acylated anthocyanins from edible sources and their applications in food systems. *Biochem. Eng. J.* **2003**, *14*, 217–225. [CrossRef]

45. Castañeda-Ovando, A.; de Lourdes Pacheco-Hernández, M.; Páez-Hernández, M.E.; Rodríguez, J.A.; Galán-Vidal, C.A. Chemical studies of anthocyanins: A review. *Food Chem.* **2009**, *113*, 859–871. [CrossRef]

46. Sadilova, E.; Stintzing, F.C.; Carle, R. Thermal degradation of acylated and nonacylated anthocyanins. *J. Food Sci.* **2006**, *71*, C504–C512. [CrossRef]

47. Patras, A.; Brunton, N.P.; O'Donnell, C.; Tiwari, B.K. Effect of thermal processing on anthocyanin stability in foods; mechanisms and kinetics of degradation. *Trends Food Sci. Technol.* **2010**, *21*, 3–11. [CrossRef]

48. Brownmiller, C.; Howard, L.R.; Prior, R.L. Processing and storage effects on monomeric anthocyanins, percent polymeric color, and antioxidant capacity of processed blueberry products. *J. Food Sci.* **2008**, *73*, H72–H79. [CrossRef] [PubMed]

49. Kalbasi, A.; Cisneros-Zevallos, L. Fractionation of monomeric and polymeric anthocyanins from concord grape (*Vitis labrusca* L.) juice by membrane ultrafiltration. *J. Agric. Food Chem.* **2007**, *55*, 7036–7042. [CrossRef] [PubMed]

50. Chung, M.Y.; Hwang, L.S.; Chiang, B.H. Concentration of perilla anthocyanins by ultrafiltration. *J. Food Sci.* **1986**, *51*, 1494–1497. [CrossRef]

51. Cissé, M.; Vaillant, F.; Pallet, D.; Dornier, M. Selecting ultrafiltration and nanofiltration membranes to concentrate anthocyanins from roselle extract (*Hibiscus sabdariffa* L.). *Food Res. Int.* **2011**, *44*, 2607–2614. [CrossRef]

52. Pap, N.; Mahosenaho, M.; Pongrácz, E.; Mikkonen, H.; Jaakkola, M.; Virtanen, V.; Myllykoski, L.; Horváth-Hovorka, Z.; Hodúr, C.; Vatai, G. Effect of ultrafiltration on anthocyanin and flavonol content of black currant juice (*Ribes nigrum* L.). *Food Bioprocess Technol.* **2012**, *5*, 921–928. [CrossRef]

53. Patil, G.; Raghavarao, K.S. Integrated membrane process for the concentration of anthocyanin. *J. Food Eng.* **2007**, *78*, 1233–1239. [CrossRef]

54. Fischer, U.A.; Carle, R.; Kammerer, D.R. Identification and quantification of phenolic compounds from pomegranate (*Punica granatum* L.) peel, mesocarp, aril and differently produced juices by HPLC-DAD–ESI/MS. *Food Chem.* **2011**, *127*, 807–821. [CrossRef] [PubMed]

55. Rwabahizi, S.; Wrolstad, R.E. Effects of mold contamination and ultrafiltration on the color stability of strawberry juice and concentrate. *J. Food Sci.* **1988**, *53*, 857–861. [CrossRef]

56. Rossi, M.; Giussani, E.; Morelli, R.; Lo Scalzo, R.; Nani, R.C.; Torreggiani, D. Effect of fruit blanching on phenolics and radical scavenging activity of highbush blueberry juice. *Food Res. Int.* **2003**, *36*, 999–1005. [CrossRef]

57. Sablani, S.S.; Andrews, P.K.; Davies, N.M.; Walters, T.; Saez, H.; Syamaladevi, R.M.; Mohekar, P.R. Effect of thermal treatments on phytochemicals in conventionally and organically grown berries. *J. Sci. Food Agric.* **2010**, *90*, 769–778. [CrossRef] [PubMed]

58. Stein-Chisholm, R.E.; Losso, J.N.; Finley, J.W.; Beaulieu, J.C. Not from concentrate blueberry juice extraction utilizing frozen fruit, heated mash and enzyme processes. *HortTechnology* **2017**, *27*, 30–36. [CrossRef]

59. Giusti, M.M.; Wrolstad, R.E. Characterization and measurement of anthocyanins by UV-visible spectroscopy. In *Current Protocols in Food Analytical Chemistry*; John Wiley & Sons, Inc.: Hoboken, NJ, USA, 2001; Available online: http://onlinelibrary.wiley.com/doi/10.1002/0471142913.faf0102s00/abstract (accessed on 22 November 2017).

60. Lee, J.; Rennaker, C.; Wrolstad, R.E. Correlation of two anthocyanin quantification methods: HPLC and spectrophotometric methods. *Food Chem.* **2008**, *110*, 782–786. [CrossRef]

61. You, Q.; Wang, B.; Chen, F.; Huang, Z.; Wang, X.; Luo, P.G. Comparison of anthocyanins and phenolics in organically and conventionally grown blueberries in selected cultivars. *Food Chem.* **2011**, *125*, 201–208. [CrossRef]

62. Wang, S.Y.; Chen, C.-T.; Sciarappa, W.; Wang, C.Y.; Camp, M.J. Fruit quality, antioxidant capacity, and flavonoid content of organically and conventionally grown blueberries. *J. Agric. Food Chem.* **2008**, *56*, 5788–5794. [CrossRef] [PubMed]

63. Nicoué, E.É.; Savard, S.; Belkacemi, K. Anthocyanins in wild blueberries of Quebec: Extraction and identification. *J. Agric. Food Chem.* **2007**, *55*, 5626–5635. [CrossRef] [PubMed]

64. Plaza, M.; Turner, C. Pressurized hot water extraction of bioactives. *Trend. Anal. Chem.* **2015**, *71*, 39–54. [CrossRef]

MDPI

St. Alban-Anlage 66

4052 Basel

Switzerland

Tel. +41 61 683 77 34

Fax +41 61 302 89 18

www.mdpi.com

Beverages Editorial Office

E-mail: beverages@mdpi.com

www.mdpi.com/journal/beverages

www.ingramcontent.com/pod-product-compliance
Lightning Source LLC
Chambersburg PA
CBHW051909210326
41597CB00033B/6083